TECHNIQUES FOR THE
ASSESSMENT OF MICROBIAL PRODUCTION
AND DECOMPOSITION IN FRESH WATERS

SOROKIN, Y I
TECHNIQUES FOR THE ASSESSMENT
002313549

HCL   QR105.5.S71

THE UNIVERSITY OF LIVERPOOL
HAROLD COHEN LIBRARY

4b

ase return or renew, on or before the latest date amped below. A fine is payable on late returned items. Books may be recalled after one week for the use of another reader.

For full conditions of borrowing, see the Library Regulations.

22 MAR 1988

IBP HANDBOOK No. 23

Techniques for the Assessment of Microbial Production and Decomposition in Fresh Waters

Edited by

Y. I. SOROKIN (USSR)

and

H. KADOTA (JAPAN)

INTERNATIONAL BIOLOGICAL PROGRAMME
7 MARYLEBONE ROAD, LONDON NW1

BLACKWELL SCIENTIFIC PUBLICATIONS
OXFORD LONDON EDINBURGH MELBOURNE

© 1972 INTERNATIONAL BIOLOGICAL PROGRAMME
and Published for them by
Blackwell Scientific Publications
Osney Mead, Oxford,
3, Nottingham Street, London W1,
9, Forrest Road, Edinburgh,
P.O. Box 9, North Balwyn, Victoria, Australia.

*All rights reserved. No part of this publication
may be reproduced, stored in a retrieval system,
or transmitted, in any form or by any means,
electronic, mechanical, photocopying, recording
or otherwise without the prior permission of
the copyright owner.*

ISBN 0 632 08850 8

FIRST PUBLISHED 1972

Distributed in the U.S.A. by
F.A. Davis Company, 1915 Arch Street,
Philadelphia, Pennsylvania

Printed in Great Britain by
BURGESS AND SON (ABINGDON) LIMITED
and bound by
THE KEMP HALL BINDERY, OXFORD

Contents

List of contributors (including observers at the Leningrad Meeting) ix

Foreword xiii

Preface xv

1   Introduction: Y. I. Sorokin and H. Kadota 1

2   **Measurement of Biological N_2 Fixation with $^{15}N_2$ and Acetylene: R. H. Burris** 3
2.1   Introduction 3
2.2   Methods 4
     2.2.1   Exposure vessels 4
     2.2.2   Sampling of phytoplankton 5
     2.2.3   Evacuation and flushing 6
     2.2.4   Adding acetylene 7
     2.2.5   Adding $^{15}N_2$ 9
     2.2.6   Incubation, inactivation and sealing 10
     2.2.7   Analysis 11
     2.2.8   Exposures to $^{15}N_2$ 12
     2.2.9   Portable field equipment and mobile laboratory ... 13
     2.2.10 Conclusion 14

3   **Measurement of Microbial Activity in Relation to Decomposition of Organic Matter** 15
3.1   Introduction: H. Kadota 15
3.2   Methods for estimating respiration rates of plankton and bacteria in natural waters: Y. Tezuka 16

3.3	Measurement of the decomposition rate by use of ^{14}C-glucose as substrate: H. Kadota...........18
3.4	Measurement of uptake of organic matter by micro-organisms: J. Overbeck 20
	3.4.1 Principle of the method 20
	3.4.2 Application of the method 22
	3.4.3 Estimation of natural substrate concentration 22
3.5	Calculation of the rate of microbial decomposition of organic matter from the heterotrophic uptake of $^{14}CO_2$: V. I. Romanenko...........23
3.6	Decomposition and mineralization of nitrogenous matter: H. L. Golterman...........23
3.7	Decomposition of organic matter in bottom sediments: S. I. Kuznetsov...........24
	3.7.1 Introduction 24
	3.7.2 Estimation of aerobic decomposition 25
	3.7.3 Estimation of the rate of anaerobic decomposition 26
3.8	Measuring the dehydrogenase activity in bottom sediments by using triphenyltetrazolium chloride: W. Ohle...........27
3.9	Measuring the evolution rate of gases in bottom sediment: W. Ohle 29
3.10	Geochemical methods for studies on the decomposition process of organic matter in natural waters: T. Koyama, N. Handa and T. Tomino...........33
	3.10.1 Mineralization rates of organic carbon and nitrogen...........33
	3.10.2 Decomposition process of carbohydrate...........36
	3.10.3 Decomposition process of proteinous compounds...........37
4	**Determination of Microbial Numbers and Biomass** 40
4.1	Sampling techniques: Y. I. Sorokin and H. W. Jannasch 40
4.2	Direct microscopic counting of micro-organisms: Y. I. Sorokin and J. Overbeck...........44
4.3	Direct count of micro-organisms in bottom sediments: Y. I. Sorokin...........47

4.4	Determination of cell size of micro-organisms for the calculation of biomass: V. Straškrabova and Y. I. Sorokin	48
4.5	Direct count using fluorescent microscopy: E. G. F. Wood	50
4.6	Use of electron microscope: S. I. Kuznetsov	51
4.7	Capillary method for the study of periphytonic organisms: T. W. Aristovskaya	53
	4.7.1 Capillary periphytonometer by Perfiliev	53
	4.7.2 Capillary pelescope by Perfiliev	57
4.8	Enumeration of viable cells of micro-organisms by plate count technique: J. W. Hopton, U. Melchiorri-Santolini and Y. I. Sorokin	59
	4.8.1 The general procedure	59
	4.8.2 The method of making dilutions	61
	4.8.3 Nature of the plating medium and the condition of incubation	63
4.9	Enumeration of microbial concentration in dilution series (MPN): U. Melchiorri-Santolini	64
4.10	Relation between cultural and microscopic counts of micro-organisms: V. Straškrabova	70
4.11	Application of different methods of enumeration of micro-organisms: Y. I. Sorokin	70
4.12	Total microbial biomass estimation by ATP method: O. Holm-Hansen	71
	4.12.1 Principle	71
	4.12.2 Outline of procedure	75
	4.12.3 Extrapolation of ATP values to biomass	75
	4.12.4 Sensitivity and precision of method	76
4.13	Evaluation of the aggregation level of planktonic bacteria: Y. I. Sorokin	76
5	**Estimation of Production Rate and *in situ* Activity of Autotrophic and Heterotrophic Micro-organisms**	**77**
5.1	Microcolony method: V. Straškrabova	77
5.2	Estimation of changes in number of bacteria in the isolated water samples: D. S. Gak, E. P. Romanova, V. I. Romanenko and Y. I. Sorokin	78

5.3	Estimation of production of heterotrophic bacteria using ^{14}C: V. I. Romanenko, J. Overbeck and Y. I. Sorokin	82
	5.3.1 Introduction	82
	5.3.2 Estimation technique	84
5.4	Calculation of rate of microbial production from rate of oxygen consumption: V. I. Romanenko and Y. I. Sorokin	86
5.5	Production of autotrophic micro-organisms: Y. I. Sorokin	86
	5.5.1 Estimation of bacterial photosynthesis	86
	5.5.2 Estimation of chemosynthesis	87
5.6	Estimation of biochemical activity of natural microflora: Y. I. Sorokin	88
5.7	The use of continuous culture: H. W. Jannasch	90
5.8	Estimation of efficiency of biosynthesis of microbial cells: Y. I. Sorokin	91
5.9	Remark on microbial production in the freshwater community: Y. I. Sorokin	92
6	**Evaluation of the Trophic Role of Micro-organisms:** Y. I. Sorokin	94
6.1	Nonisotopic methods	94
6.2	Isotopic methods	95
	References	99
	Index	106

List of contributors (including observers at the Leningrad meeting)

T. W. Aristovskaja, Soil Museum, Laboratory of Soil Microbiology, Birzhevoi Projesd 6, Leningrad, USSR.

O. N. Bauer, Zoological Institute, USSR Academy of Sciences, Leningrad W-164, USSR.

R. H. Burris, Department of Biochemistry, University of Wisconsin, Madison, Wisconsin 53706, USA.

V. G. Drabcova, Laboratory of Limnology, Petrovskaya 3A, Leningrad, USSR.

D. Z. Gak, Hydrobiological Institute, Academy of Sciences of the Ukraine, Vladimirskaja 44, Kiev, USSR.

M. Gerletti, Istituto Italiano di Idrobiologia, 28048 Verbania, Pallanza, Novara, Italia.

A. B. Getzova, Zoological Institute, USSR Academy of Sciences, Leningrad W-164, USSR.

H. L. Golterman, Limnological Institute, 'Vijverhof', Nieuwersluis, Holland.

K. V. Gorbunov, Astrakhan Technical Institute of Fish Industry and Fishery, Tatistcheva 16, Astrakhan, USSR.

B. L. Gutelmacher, Laboratory of Limnology, Zoological Institute, USSR Academy of Sciences, Leningrad W-164, USSR.

O. Holm-Hansen, Institute of Marine Resources, University of California, La Jolla, California 92037, USA.

J. W. Hopton, Department of Microbiology, the University of Birmingham, Birmingham 15, England.

H. W. Jannasch, Woods Hole Oceanographic Institution, Woods Hole, Massachusetts 02543, USA.

List of contributors

H. Kadota, Laboratory of Microbiology, Research Institute for Food Science, Kyoto University, Kyoto, Japan.

T. Koyama, Water Research Laboratory, Faculty of Science, Nagoya University, Chikusa-ku, Nagoya, Japan.

O. M. Kozhova, University of Irkutsk, USSR.

S. I. Kuznetsov, Institute of Microbiology, USSR Academy of Sciences, Profsojusnaja 7, Moscow, USSR.

M. N. Lebedeva, Institute of Biology of the South Seas, Sevastopol, USSR.

U. Melchiorri-Santolini, Istituto Italiano di Idrobiologia, 28048 Verbania, Pallanza, Novara, Italia.

W. Ohle, Max-Planck-Institut für Limnologie, 232 Plön, Holstein, German Federal Republic.

J. Overbeck, Max-Planck-Institut für Limnologie, 232 Plön, Holstein, German Federal Republic.

L. N. Pshenin, Institute of Biology of the South Seas, Nachimova 2, Sevastopol, USSR.

V. I. Romanenko, Institute of Inland Water Biology, USSR Academy of Sciences, Borok, USSR.

A. P. Romanova, Institute of Fish Culture in Lakes and Rivers, Makarova Quay 26, Leningrad, USSR.

J. Rzóska, IBP Central Office, 7 Marylebone Road, London, N.W.1, England.

N. N. Smirnov, Institute of Evolutionary Morphology and Ecology of Animals, USSR Academy of Sciences, Leninprospekt 33, Moscow, USSR.

S. Sobot, Institut za Oceanografiju i Ribarstvu, Split, Yugoslavia.

Y. I. Sorokin, Institute of Inland Water Biology, USSR Academy of Sciences, Borok, USSR.

V. Straškrabova, Hydrobiological Laboratory, Czechoslovakian Academy of Sciences, Prague 2, CSSR.

Y. Tezuka, Faculty of Science, Tokyo Metropolitan University, Setagaya-ku, Tokyo, Japan.

G. G. Winberg, Laboratory of Freshwater Biology, Zoological Institute, USSR Academy of Sciences, Leningrad W-164, USSR.

E. J. F. Wood, Institute of Marine Science, University of Miami, Florida 33149, USA.

V. E. Zaika, Institute of Biology of the South Seas, Academy of Sciences of Ukr. SSR, Sevastopol, USSR.

T. V. Zharova, Zoological Institute, USSR Academy of Sciences, Leningrad W-164, USSR.

Foreword

This is the sixth handbook to be issued by Section PF (Production in Freshwaters) of the IBP.*

Previous handbooks are: *Handbook No. 3*—Methods for Assessment of Fish Production in Freshwaters—edited by W. E. Ricker (1968), now appearing in a considerably revised edition (1971). *Handbook No. 8*—Methods for Chemical Analysis of Freshwaters—edited by H. Golterman with the assistance of R. S. Clymo (1968, 1970, 1971). *Handbook No. 12*—A Manual of Methods of Measuring Primary Productivity in Aquatic Environments—edited by R. A. Vollenweider (1969, 1971). *Handbook No. 17*—A Manual of Methods for the Assessment of Secondary Productivity in Freshwaters—edited by W. T. Edmondson and G. G. Winberg (1971). *Handbook No. 21*—Project Aqua, A Source Book of Inland Waters Proposed for Conservation—edited by H. Luther and J. Rzóska (1971).

The present volume is an attempt to bring together the main methods of microbial assessment. The role of micro-organisms in the biological functioning of a water body is fundamental through chemo- and photo-synthesis and decomposition and has a profound effect on the circulation of nutrients. Their role was recognized a long time ago but only recently was defined with some precision. Without the assessment of the role of microbial organisms, the complexity of production cannot be grasped fully. It is, therefore, gratifying that, within the IBP, an attempt has been made to collect and present the available methods of research, even though some of them have not yet reached finality.

This was a difficult task undertaken by a number of microbiologists from eight countries during a working meeting in Leningrad (1969) and subsequently by meetings of the editors—Dr. Y. Sorokin, chief biologist at the Institute of Inland Waters of the U.S.S.R. Academy of Sciences at Borok; Professor H. Kadota, working at the Research Institute for Food Science at

* The International Biological Programme is a worldwide plan concerned with 'the biological basis of productivity and human welfare'.

Kyoto University. Dr. H. Jannasch (Woods Hole, U.S.A.) and Dr. J. Hopton (Birmingham University U.K.) have helped with advice.

We are most grateful to the Editors for fulfilling their difficult task.

The interest and financial help shown by UNESCO in this endeavour is acknowledged with much appreciation.

October 1971

E. B. WORTHINGTON
Scientific Director
IBP Central Office
7 Marylebone Road
London, N.W.1 5HB

Preface

The International Biological Programme concentrated its efforts on a worldwide appraisal of the biological productivity of terrestrial and aquatic environments in relation to human welfare. Freshwater, as a separate environmental entity, has been dealt with in four preceding meetings on 'primary', 'secondary' productivity (invertebrates and fish) and the chemical environment; the corresponding publications have appeared in the form of IBP/PF Handbooks. The present treatise is concerned with microbial processes that are usually not covered in the classical limnological aspects of productivity. Methods employed in the assessment of microbial activities are, in general, remarkably different from those used in studies on plant and animal populations.

The description of methods in this book is based on contributions and discussions during the IBP/PF technical meeting on microbial production and decomposition held under the sponsorship of the Academy of Sciences of the USSR, in Leningrad on May 27–31, 1969. In this meeting, 35 aquatic microbiologists from 8 countries discussed the present status of our knowledge in microbial activities in fresh waters and methods for their quantitative assessment. In order to make the book, as far as possible, into a coherent entity, the editors found it necessary to exercise their rights. Some individual contributions are, therefore, printed with little change but others have undergone changes so as to fit them into the general pattern and still others appear only in joint chapters. The meeting was held in five sections. The title and convener of each section were:

1. Measurement of nitrogen fixation in aquatic environments (Convener, R. H. Burris).
2. Measurement of microbial decomposition of organic matter (Convener, II. Kadota).
3. Estimation of cell number and biomass of micro-organisms (Convener, V. Straškrabova).

4. Estimation of production rate of micro-organisms (Convener, Y. I. Sorokin).
5. Evaluation of the trophic role of micro-organisms (Convener, Y. I. Sorokin).

Most methods in ecology of micro-organisms in fresh waters are new or in the process of being developed. Therefore, this first attempt to collect techniques in such a new field of science obviously suffers from incompleteness due to the limited number of participants and from unevenness of scientific style. As limnology is advancing rapidly, inevitably the methods described in this book will be modified and improved.

We are indebted by Dr. Julian Rzóska, Scientific Coordinator of IBP/PF for his constant help throughout the meeting and in editing this book. We are also grateful to Professors G. G. Winberg and O. N. Bauer of the Academy of Sciences of the USSR for their kind hospitality during the meeting. We are also very grateful to Dr. H. W. Jannasch for his kind help in editing some of the manuscripts. Dr. J. W. Hopton, of the University of Birmingham, has read the script critically as to style.

We acknowledge with gratitude the considerable help for our task from UNESCO.

Yuri I. Sorokin,
Institute of Inland Water Biology,
USSR Academy of Sciences,
Borok, Nekouz,
Jaroslav,
USSR.

Hajime Kadota,
Research Institute for Food Science,
Kyoto University,
Kyoto,
Japan.

1
Introduction

The role of micro-organisms in biological production in aquatic environments is complex and difficult to establish theoretically as well as methodologically.

It has been a major problem in aquatic microbiology to establish a common terminological basis for describing limnological and bacteriological processes in order to facilitate effective cooperation between hydrobiologists and microbiologists. The term 'microbial production' has been chosen for practical reasons and requires a definition.

In contrast to plants and animals, micro-organisms are not restricted to a single metabolic type but include various groups of photosynthetic, chemosynthetic, and heterotrophic organisms. Consequently, microbial production consists of primary and secondary production at the same time.

Compared to the extensive primary production by phytoplankton in fresh water, photosynthetic activity of micro-organisms is restricted to a limited area and very specific environmental conditions. Photosynthetic microbial production is of importance only in environments where light energy and appropriate electron donors such as H_2S, are simultaneously available.

Micro-organisms often enhance the productivity of water bodies by making available to organisms living there, organic matter originally produced in the surrounding areas and transported into the water body by the movement of water. Thus, in some aquatic ecosystems, besides the primary production by photosynthesis, microbial production at the expense of allochthonous organic matter cannot be neglected as a contribution to the food chain. In some water bodies, which have large surrounding drainage basins, the production of micro-organisms at the expense of energy of allochthonous materials from land or from other water bodies can be of the same order of magnitude as autochthonous primary production by plants and can sometimes exceed it.

Secondary production by heterotrophic micro-organisms will be of great quantitative importance in most situations. Compared to the secondary production by animals, microbial secondary production is of special importance for two reasons: (1) micro-organisms are capable of attacking organic substrates that cannot be utilized by animals, and (2) micro-organisms produce particulate food materials from dissolved organic materials and, therefore, represent an important link in the natural food chain.

During secondary production, decomposition processes release the energy necessary for biosynthesis, and release also mineralized nutrients for primary production. For these two important reasons, microbial production cannot be separated from microbial decomposition, neither in theoretical treatments nor from practical considerations.

From the above considerations it is quite obvious that micro-organisms are of importance in the processes of mineralization and nutrient regeneration as well as in the creation of the basic food resources in the aquatic environment. Therefore the development of the methods of evaluation of the microbial production and decomposition is now extremely important in the study of ecosystems in fresh waters.

Microbial processes of nitrification, denitrification, oxidation of inorganic sulfur compounds, and sulfate reduction are also of importance for the evaluation of biological productivity. But these transformations are not specifically treated in this text.

Besides bacteria, other micro-organisms such as moulds, yeasts, streptomyces and viruses can play an important role in some aquatic ecosystems. These organisms, however, are also not treated, since information about them from an ecological point of view is still limited.

2

Measurement of Biological N_2 Fixation with $^{15}N_2$ and Acetylene

Although the importance of biological N_2 fixation is obvious in the agricultural economy and in aquatic systems, the methods for its quantitation have been so deficient that its adequate evaluation has never been possible. The reduction of acetylene to ethylene can now serve as an index of N_2 fixation to furnish quantitative measurements of N_2 fixation in the field.

2.1 Introduction

Schöllhorn and Burris (1966) reported that azide and acetylene were reduced by the N_2-fixing enzyme complex. Independently, Dilworth (1966) also found that acetylene was reduced and demonstrated that ethylene was the product of the reduction.

An extensive application of the acetylene reduction method for field studies was reported by Stewart, Fitzgerald and Burris (1967) who employed the method to examine N_2 fixation in soil, in excised nodules from leguminous and non-leguminous plants, and in blue-green algae in lakes. In 1968 Hardy *et al.* reported that they also had used the method for investigation of N_2 fixation in soils and in leguminous nodules; they employed syringes as their vessels for exposure of samples.

The acetylene reduction method for measuring N_2 fixation in aqueous environments and in the soil by free-living and symbiotic systems is particularly attractive, because it is simple, cheap, and extremely sensitive. The opinion is generally held that a measurement of acetylene reduction to ethylene can be employed as a valid index of N_2 fixation based upon the observations that: (1) the reduction of acetylene, like the reduction of N_2, requires ATP and a reducing agent such as dithionite or reduced ferredoxin; (2) as the enzyme system is purified for N_2 fixation it is purified for C_2H_2 reduction in a parallel fashion; (3) inactivation of N_2-fixing capacity is accompanied by inactivation of C_2H_2-reducing capacity; (4) the Fe protein and the FeMo protein are required in combination both for N_2 and C_2H_2 reduction, as neither protein functions alone and (5) C_2H_2 inhibits N_2 fixation.

Although the evidence for the validity of C_2H_2 reduction as an index of N_2 fixation is strong, the method must be applied with caution, because it is an indirect method. It is inexcusable to use such an indirect method as the sole index of N_2 fixation in the laboratory, where measurements of NH_3 formation or $^{15}N_2$ fixation can be made rather easily. Until its validity and the ratio of C_2H_2 to N_2 reduction is more clearly established, its use in the field should also be accompanied by periodic control tests of $^{15}N_2$ which constitute a direct and absolute test for N_2 fixation.

2.2 Methods

It is apparent that there are a variety of ways in which one can employ C_2H_2 reduction as an index of N_2 fixation. The methods to be described have been employed successfully, but they doubtless can be modified to improve them for specific applications. In general, they have the advantage of simplicity and they employ relatively inexpensive equipment.

2.2.1 Exposure vessels. Vaccine bottles (designed for serum stoppers without a sleeve as closures) or serum bottles (designed for sleeve-type serum stoppers) are cheap, rugged, designed to carry gas-tight serum stoppers and available in a variety of sizes (For example in the USA, A. H. Thomas Co., Philadelphia, Pa. 19105, USA, stocks serum bottles in nominal sizes of 5, 10, 20, 30, 50, 100 and 125 ml and vaccine bottles in nominal sizes of 15, 30, 60 and 100 ml; the actual sizes are somewhat greater.) For homogeneous samples the small sizes are suitable, but for non-homogeneous material the larger bottles providing for larger samples are advised.

Hardy et al. (1968) have suggested the use of 50 ml plastic hypodermic syringes as exposure vessels, and they have employed these successfully with soil and root nodule samples. The units are somewhat less convenient to handle, specially with liquid samples, are more susceptible to leakage than rubber stoppered bottles, and there is no apparent advantage in gassing them. Their two advantages are (a) with the plunger removed they offer a large orifice for the introduction of soil cores and other materials, and (b) the gas is removed from above the reactants at the end of the reaction thus precluding the necessity for inactivating the biological agent to terminate the reaction.

The research worker should try both serum bottles and syringes to see which fits his particular needs better. A combination of the two methods may be the best choice; *i.e.*, the biological material is gassed and exposed in a serum bottle, and at the termination of the experiment a sample of gas is removed into a disposable hypodermic syringe while displacing fluid is added to the reaction bottle. The needle may be capped (insert into a piece of rubber) for storage until the gas can be removed from the syringe for injection into the gas chromatographic (GC) apparatus. Conventional syringes with glass plungers are not gas tight, but disposable syringes with plungers carrying rubber or plastic rings are quite reliable for limited periods. They also are entirely adequate for injecting gas samples into the GC apparatus. Disposable syringes with glass barrel and plastic plunger appear to be best, but the all-plastic syringes are adequate (A. H. Thomas carried B-D Plastipak disposable syringes in 1, 2·5, 5 and 10 ml sizes).

2.2.2 Sampling of phytoplankton*. Heavy algal growths can be measured directly for C_2H_2 reduction without concentration, but under most conditions the phytoplankton must be concentrated. The N_2-fixing blue-green algae occur predominantly in natural habitats as filamentous forms. Filaments can be recovered effectively with a fine-mesh plankton net (about 150 meshes per 2·54cm —1 in). Bacteria and unicellular algae must be recovered with membrane filters. Filtration may damage some delicate colonial algae, and controls to examine this possibility should be devised.

The water sample of known volume taken with a sampling bottle is passed through plankton netting fastened with contact cement to a Bunsen burner tripod. If the tripod is placed in a pail or pan the filtrate can be captured to be discarded or used as wash or diluent. The phytoplankton mass is washed to the middle of the plankton net with filtered water delivered from a wash bottle. This concentrated suspension is transferred with a Pasteur pipette and rubber dropper bulb to a 5 ml graduated cylinder; the plankton net is rinsed with filtered water 3 or 4 times to complete the transfer and to give a total volume of algal suspension of about 4 ml (the volume is recorded). The algae are suspended uniformly by sucking them into and expelling them from a 1 ml

---

* For more detailed descriptions of phytoplankton methods see IBP Handbook No. 12: R. A. Vollenweider (edit.), A Manual of Methods for Measuring Primary Production in Aquatic Environments.

volumetric pipette, and then 1 ml samples in triplicate are transferred to serum bottles of 5 ml nominal and about 7 ml actual size.

In waters with a low population of N_2-fixing phytoplankton it may be necessary to concentrate the phytoplankton further. The water sampling and filtration are performed as before, but instead of re-suspending the algae in a graduated cylinder they are transferred from the plankton net directly to a small Hirsch funnel. Again the plankton net must be rinsed with filtered water 3 or 4 times to aid in completing the transfer. The Hirsch funnel is of a size to take a 12—15 mm disc of filter paper. The funnel is inserted into a rubber bulb (approximately 60 ml—2 ounces capacity), the bulb is squeezed, the filter paper is moistened and seated, and the phytoplankton suspension is added with the Pasteur pipette to the filter paper while the bulb is slowly released to create a vacuum to speed filtration. The entire filter paper with its phytoplankton is transferred to a 5 ml serum bottle, 1 ml of filtered water is added and the bottle is stoppered preparatory for treatment. It is apparent that the density of the phytoplankton will govern the size of sample concentrated for exposure.

2.2.3 Evacuation and flushing. Air may be replaced in a vessel either by evacuating it or by removing it by flushing with another gas. Both methods are applicable for the acetylene reduction assay. When an electricity supply is available, the use of an ordinary laboratory vacuum pump with rotating vanes operating in oil is desirable. Addition of a handle to the pulley of such a pump permits its effective hand operation in the field. The more compact laboratory pumps with 'dry' rotating vanes seldom give less than 0·2 atm. residual pressure and must be used at least 3 times with intermediate flushing to attain adequate gas displacement (residual gas after first evacuation 0·2 atm., second 0·04 atm., third 0·008 atm.).

Flushing can be accomplished with a pressurized cylinder of gas (the mixture consisting of 21% O_2, 0·04% CO_2 and 78·96% argon is usually used, but other mixtures including mixtures for anaerobic incubation also are suitable) equipped with a pressure-reducing valve. Gas is passed into the reaction bottle through a hypodermic needle inserted through the septum of the rubber serum stopper and is vented through a second needle inserted through the septum. Vigorous gassing for a minute is adequate.

A manifold carrying about 4 hypodermic needles is convenient for evacuating and gassing samples in groups. In replacement of gas by flushing, it is difficult to judge whether all vessels on a manifold are being purged equally, so it is desirable to flush individual samples rather than to use a manifold.

Hardy et al. (1968) have described the displacement of gas and addition of gas when syringes are used as exposure vessels.

2.2.4 Adding acetylene. Evacuation or flushing will have removed air and replaced it with the desired gas mixture minus acetylene or $^{15}N_2$. The reaction now is initiated by injecting approximately 0·1 atm. C_2H_2 into the reaction vessel. This is accomplished with a hypodermic syringe. For example, vessels with approximately 5 ml gas volume receive an injection of 0·5 ml acetylene to initiate the reaction at time zero. Thus, samples during exposure are at a pressure above atmospheric pressure, and any small leakage is outward.

It is not obligatory that samples be evacuated or flushed before adding C_2H_2, although with root nodules and other large pieces of tissue it may be desirable. The affinity of the N_2-fixing enzyme system for C_2H_2 is sufficiently high so that if 0·2 atmosphere of C_2H_2 is injected into a bottle filled with air, the N_2 does not compete effectively with the C_2H_2 and reduction of C_2H_2 is virtually the same as in the absence of N_2. This is a great convenience for field work, as avoiding evacuation or flushing simplifies the operations substantially. After adding the 0·2 atmosphere of C_2H_2 we vent excess gas by inserting a needle through the septum for a few seconds. The investigator may compare his samples with and without evacuation or flushing to establish that his material does not require these operations.

Acetylene can be taken from an acetylene generator, a pressurized cylinder, a displacing bottle or a rubber bladder filled with the gas. Acetylene cylinders contain acetone, but the amount of acetone carried in the gas is slightly or non-inhibitory to N_2 fixation, so we usually do not remove it. If acetone is considered undesirable, it can be removed by passing the gas through a dry ice trap or through a concentrated sulfuric acid trap. If heavy rubber tubing is placed on the outlet from the gas cylinder, the tubing can be filled with C_2H_2, its end can be plugged, and the C_2H_2 then can be removed through the tubing with a hypodermic needle and syringe. A convenient storage bottle for C_2H_2 can be made as shown in Fig. 1. It is filled as follows.

Fill reservoir C through port A with boiled water. Insert an open hypodermic needle in stopper B. Connect tubing terminated with a hypodermic needle to the cylinder of C_2H_2 and flush the tubing. Insert the needle in stopper A and add gas to displace water from C into D. Remove both needles. Store the reservoir in an upright position. When you need C_2H_2,

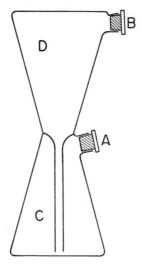

Figure 1. Reservoir for C_2H_2.

insert an open needle into stopper B and remove gas through stopper A with a hypodermic needle and syringe.

G. A. Fitzgerald has suggested a particularly convenient C_2H_2 reservoir for field work (Fig. 2). Obtain the rubber bladder for a football, volleyball or basketball and the needle used for its inflation. Cement the inflation needle

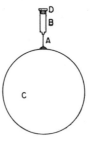

Figure 2. Bladder serving as reservoir for C_2H_2.

A to a 5 ml glass barrel of a hypodermic syringe B with epoxy cement. Insert the inflating needle adaptor (moisten the needle for insertion), and collapse the bladder C to remove air. Remove the needle adaptor and attach it with rubber tubing or rubber tubing and a stopper to a cylinder of C_2H_2. Purge the tubing and adaptor with C_2H_2, and then insert the inflation needle into the bladder and inflate it with C_2H_2. Remove the inflation needle for transporting the bladder. To remove C_2H_2 from the bladder, fill the needle adaptor reservoir with water, close the open top with a serum stopper D, and insert the needle adaptor into the bladder. Insert a hypodermic needle through the serum stopper, and displace the water from the needle adaptor reservoir through the hypodermic needle. As needed, withdraw C_2H_2 through the serum stopper with a hypodermic needle and syringe. The bladder can be held conveniently between your legs.

Fig. 3 illustrates a simple generator for C_2H_2 which can be used in the field. Blow small holes near the bottom of test tube B and constrict it with indents

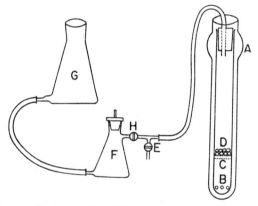

Figure 3. Apparatus for generation of C_2H_2 in the field. The entire unit can be mounted on a board or metal stand. Design adapted by Timothy Mague.

or seal in a coarse sintered disc C as shown to hold lumps of CaC_2. Add the CaC_2 lumps D and stopper the tube. Add water in test tube A to a level slightly above the CaC_2. The water will moisten the CaC_2 and C_2H_2 will be generated. Allow the generated C_2H_2 to sweep air from the system through E. Close E, open H, and then collect C_2H_2 in the reservoir F by displacing boiled water from F into reservoir G. When the stopcock H is closed, the pressure developed will force water from B back into tube A, and generation of C_2H_2 will cease. Withdraw C_2H_2 through the serum stopper I.

2.2.5 Adding $^{15}N_2$. When ^{15}N is to be used as a tracer it is necessary to remove N_2 from the system by evacuation or flushing as described earlier.

Then 0·2 atm. of $^{15}N_2$ is injected with a hypodermic syringe and needle. Because $^{15}N_2$ is costly, it is advantageous to use small reaction vessels. We usually use serum bottles with about 7 ml volume (nominal 5 ml bottles) containing 1 ml of suspension and inject 1·2 ml of $^{15}N_2$ (30—95 atom % ^{15}N excess) to give about 0·2 atm. $^{15}N_2$. Controls should be run in the laboratory to establish whether N_2 fixation at 0·8 atm. N_2 is detectably greater than at 0·2 atm. N_2.

The $^{15}N_2$ gas can be stored in a reservoir as shown in Fig. 4. As $^{15}N_2$ supplied commercially is commonly contaminated with nitrogen oxides, it should be

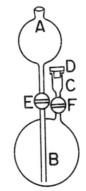

Figure 4. Reservoir for $^{15}N_2$.

transferred to a bulb containing alkaline permanganate (50 g $KMnO_4$ and 25 g KOH in 1 litre H_2O) to oxidize the compounds to higher oxides which are then absorbed by the alkali. The gas then can be transferred with a Toepler pump or other transfer device to the evacuated reservoir shown. Displacing fluid (20% $Na_2S_2O_4$ in 5% H_2SO_4) is added from A to B to bring the gas in B to atmospheric pressure plus the hydrostatic head pressure. The acid displacing fluid absorbs any residual NH_3 that may be in the $^{15}N_2$. Small reservoir C is evacuated through a hypodermic needle inserted through serum stopper D. The needle is removed and C is filled with gas from B. To remove $^{15}N_2$ for injection into reaction bottles, open stopcocks E and F, insert a hypodermic needle through D and withdraw $^{15}N_2$ into the attached syringe.

2.2.6 Incubation, inactivation and sealing. Samples can be incubated under *in situ* conditions or standard conditions in the field or on the deck.

Samples of phytoplankton or other 'homogeneous' agents can be inactivated conveniently by injecting a volume of 5N H_2SO_4 approximately equivalent to 0·2 the volume of the reaction mixture; the sample is shaken and the acid apparently inactivates the phytoplankton immediately. Although the reaction vessels have an elevated pressure, and this reduces the opportunity for inward leakage, the stoppers should be dried and sealed at the punctured area with RTV (General Electric self-vulcanizing silicone rubber compound, RTV-112, white, pourable; other silicone rubber sealants also are suitable) after the injection of acid. The RTV sets adequately in 60 minutes and provides a barrier which can be penetrated with a hypodermic needle.

2.2.7 Analysis. For analysis, samples of gas are removed through the rubber serum stopper directly from the reaction vessel (inject 0·5 ml boiled-distilled water if samples are at atmospheric pressure). Disposable glass-barrel hypodermic syringes (0·5 to 1·0 ml) equipped with 25 or 27 gauge needles are suitable. The 0·5 ml gas sample is injected directly through the septum of the GC apparatus for analysis. A large rubber bulb can be used to sweep air through the open syringe barrel between each sample to avoid cross-contamination between samples.

A hydrogen flame ionization detector is suitable for hydrocarbon analysis, and a relatively simple GC apparatus can be employed. The use of a 2—3 meter, 3·1 mm (⅛ in) diameter column of Porapak R (80 to 100 mesh) gives good separations of acetylene and ethylene. The column can be operated at room temperature to 50° C with about 6·8 kg (15 lb) pressure (10 ml/min) of H_2 and 40 lb pressure (20 ml/min) of N_2 as the carrier gas. Under these conditions, separations are achieved in 2 minutes. A 1·2 m (4 ft), 3·1 mm (⅛ in) diameter column of Porapak N operated at 75° C gives an excellent separation of acetylene and ethylene. (The Porapaks are porous polymers in bead form of ethylvinyl-styrene or ethylvinylbenzene-divinylbenzene modified to alter their polarity.) As an alternative one may employ a 0·6 m (2 ft), 3 mm (⅛ in) diameter column of alumina (all particles smaller than 100 mesh removed) operated at 100° C. Flow rates of gases should be kept constant, and if analyses are to be made on successive days it is advisable to keep the instrument running.

Peaks are narrow and sharp, so adequate quantitation can be achieved by measuring peak heights under standardized conditions. Controls inactivated with acid before addition of C_2H_2 always are included, so that correction can be made for any contaminating C_2H_4. Standard curves (log log plot) are constructed from the analysis of carefully prepared mixtures of C_2H_2 and C_2H_4. Data can be reduced to an absolute basis by reference to the standard curves. Although the sensitivity can be pushed farther, it is desirable to measure samples with greater than 50 picomoles of C_2H_4 to achieve accuracy. Replicate samples from a single bottle normally are analysed (it is advisable to inject 0·5 ml of distilled water before each 0·5 ml sample of gas is withdrawn); these replicates should give very consistent results. There may be some variation between replicate bottles, so it is advisable to employ triplicates of each treatment when feasible.

2.2.8 Exposures to $^{15}N_2$. As indicated, exposures to $^{15}N_2$ can be made in the same way as exposures to C_2H_2. Inactivation can follow the same procedure. To determine the ^{15}N concentration of the N_2 in individual bottles, inject 1 ml of boiled distilled water and then remove 1 ml of gas from the bottle with a hypodermic syringe (27 gauge needle) and inject it into the mass spectrometer. A simple device as shown in Fig. 5 serves as a suitable adaptor for the

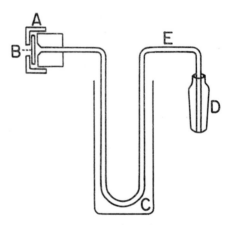

Figure 5. Adapter for introducing gas samples into a mass spectrometer.

mass spectrometer. A cap A screwed in place holds a GC rubber septum B. A small hole in the cap A permits insertion of a hypodermic needle through the septum. The injected gas passes through a liquid nitrogen or dry ice freezing trap C to remove water vapor. The unit is connected to the mass spectrometer through a standard taper joint D. The 3·1 mm ($\frac{1}{8}$ in) diameter copper tubing E is sealed inside standard taper stopper D with RTV.

As only 0·2 atm. of N_2 normally is present in the gas mixture, pressure behind the capillary leak of the mass spectrometer must be greater than that used for pure samples of N_2. Because contamination with air will produce a nonequilibrium distribution of masses 28, 29 and 30, it is not permissible to measure only masses 28 and 29 to determine the concentration of ^{15}N. Masses 28, 29 and 30 should be measured with the mass spectrometer, and from these measurements the atom % ^{15}N excess of the gas mixture can be calculated. (Example: mass 28, 17 volts peak; mass 29, 22 volts, mass 30, 10 volts. ^{14}N contribution 17 volts from 28 and 11 volts from 29; ^{15}N contribution 11 volts from 29, 10 volts from 30. Voltage from ^{15}N totals 21 volts, and total voltage of 3 peaks is 49 volts. $\frac{21}{49} \times 100 = 42·86$ atom % ^{15}N.)

The entire reaction mixture now is transferred from the reaction vessel to a Kjeldahl flask and is digested. (Add 1·5 g K_2SO_4, 1·5 ml of a solution made by dissolving 10 g red mercuric oxide in 12 ml conc. H_2SO_4 and diluting to 100 ml with H_2O, 3 ml of conc. H_2SO_4 and 2 beads; heat to digest.) After complete digestion, dilute, cool, add 0·6 g ammonia-free Zn dust, make alkaline with 10 ml 13 N NaOH and distil the NH_3 into 10 ml of 0·1 N H_2SO_4. Adjust the distillate to 50 ml, nesslerize a 1/50 to 1/100 aliquot to determine total N and concentrate the main sample for conversion to N_2 with alkaline hypobromite (Burris and Wilson, 1957). Then analyse the N_2 sample for ^{15}N with a mass spectrometer.

It is important to run ^{15}N samples periodically to establish a valid ratio of C_2H_2 reduced to N_2 reduced. Otherwise one has no sound basis for translating C_2H_2 reduction into terms of N_2 fixation. It also is advisable to examine the time course of N_2 reduction vs C_2H_2 reduction, to establish any discrepancy in their rates.

2.2.9 Portable field equipment and mobile laboratory. A small cylinder ('lecture bottle') of flushing gas, needles and syringes, reaction vessels with stoppers, a bottle (Fig. 1) or bladder (Fig. 2) of C_2H_2, a tube of RTV, and inactivating acid will serve as minimal equipment for field tests of C_2H_2 reduction. The flushing gas can be omitted if 0·2 atm. C_2H_2 is added to samples, and C_2H_2 can be generated in the field from calcium carbide plus water. The Department of Biochemistry, University of Wisconsin has equipped a ¾-ton pick-up truck with a 'camper' body as a mobile laboratory. It carries a gas chromatograph, strip chart recorder, cylinders of N_2, H_2, C_2H_2, and a purging mixture (21% O_2, 0·04% CO_2, 78·96% A), vacuum pumps, gassing manifold with aneroid pressure-vacuum gauge, torsion balance, battery operated colorimeter, pH meter, Kjeldahl digestor (gas), gasoline motor generator for 115 v. 60 cycle A.C., 6 and 12 v. batteries, boat, outboard motor, water sampling equipment, refrigerator, cooking facilities, water and gas supplies, space heater, and sleeping quarters for 4. The unit permits good mobility for field work, as well as rapid preparation and complete processing of samples. Units from the mobile laboratory can be removed for use on board a motor launch employed in sampling on large bodies of water.

2.2.10 Conclusion. The gas chromatographic analysis of C_2H_4 produced by reduction of C_2H_2 furnishes a simple, cheap and highly sensitive method for determining the N_2-fixing capacity of soils, root nodules, free-living microorganisms, enzyme extracts and phytoplankton. The method can be applied readily in the field. Although there are many indications that the method gives a valid measure of N_2 fixation, the agents being tested should be exposed periodically both to C_2H_2 and to $^{15}N_2$ to establish the relative rates of C_2H_2 and N_2 reduction.

3

Measurement of Microbial Activity in Relation to Decomposition of Organic Matter

3.1 Introduction

The greater part of organic substances produced by plants and animals in lakes and rivers and those brought there from the outside are sooner or later decomposed and mineralized through the metabolism of various aquatic organisms. Especially, the micro-organisms such as bacteria, yeasts and moulds are known to play an important role in the processes involved in the decomposition and mineralization of organic matter in aquatic environments. The activities of micro-organisms in this respect, therefore, have direct or indirect influence on the productivity of aquatic organisms of various trophic levels.

The decomposition processes in which micro-organisms participate are divided from the chemical point of view, into two categories: (1) hydrolytic breakdown of the organic high polymers which constitute the major parts of plant and animal bodies into compounds of low molecular weight, and (2) non-hydrolytic breakdown of the resulting small organic molecules, generally accompanied by the consumption of oxygen. Through the latter process (mineralization) organic molecules in aquatic environments are converted into inorganic compounds utilized as plant nutrients. The processes of decomposition and mineralization are diagrammatically shown in Fig. 6. In

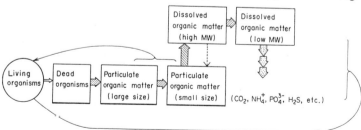

Figure 6. Decomposition of organic matter in aquatic environments.

this figure, the shadowed and dotted arrows indicate the decomposition processes in which the micro-organisms are principally participating.

To analyse the mechanism of decomposition and mineralization of organic matter in inland waters and to determine the rate of the reactions involved, either a biochemical or geochemical method is usually employed. The biochemical method involves the *in situ* estimation of the activities of micro-organisms in waters and sediments by use of the oxygen consumption methods or tracer techniques, and the geochemical one is concerned with the chemical analysis of waters and sediments. The chemical properties of waters and sediments generally reflect the biochemical events which have taken place as a consequence of microbial activities in those environments.

3.2 Methods for estimating respiration rates of plankton and bacteria in natural waters

In eutrophic freshwaters, the overall respiration rate of plankton can be determined by incubating water sample for a definite time in the dark and by measuring the initial and final dissolved oxygen concentrations. The oxygen consumption thus measured is that attributed to all planktonic organisms existing in that water sample. The separate respiration rates of plankton and bacteria can be estimated roughly. Since bacteria are smaller than most of the other planktonic organisms, the main part of the bacterial population can be separated from the other planktons by size fractionation. For this purpose, the following procedure can be employed.

An appropriate volume of water sample is collected from a required layer of a water. In order to exclude the phyto and zooplankton, one portion of the water sample is filtered immediately through a membrane filter of 8 or 5 µm porosity and the filtrate poured into several sterilized oxygen bottles. Another portion of the water sample is also poured into sterilized oxygen bottles without filtration. The initial number of bacteria and the amount of dissolved oxygen in one bottle of each series are determined. The remaining bottles are incubated in the dark *in situ*, or at a given temperature, for 24 to 48 hrs, and then the final bacterial numbers and dissolved oxygen are determined.

The number of total bacteria can best be determined by the direct microscopic count on a membrane filter. However, when the number of viable heterotrophic bacteria is desired, it can be obtained by the agar plate method using appropriate media for aquatic bacteria. Suitable ranges of incubation temperature and incubation period are: 20—30° C and 7—10 days respectively. Dissolved oxygen concentration is determined either by Winkler's method or by means of oxygen electrodes.

Calculation of respiration rate of bacteria

If the bacterial population in the filtrate is assumed to grow exponentially during t hours, the oxygen uptake per bacterial cell per hour (Qo_2) can be calculated according to the following equation (Buchanan and Fulmer, 1930):

$$Qo_2 = \frac{2.303 \, Rt^1 \log \frac{Nt^1}{No^1}}{(Nt^1 - No^1) \, t}$$

where No^1 and Nt^1 are the initial and final numbers of bacteria respectively, and Rt^1 is the amount of oxygen consumed by the bacteria in the filtrate during t hours.

If the size of the bacterial population in the filtrate is almost constant during the incubation time, Qo_2 is given by the following equation:

$$Qo_2 = \frac{2Rt^1}{(No^1 + Nt^1) \, t}$$

Assuming that the respiratory activity of one bacterial cell in the natural water is equal to that of one bacterial cell in the filtrate, the oxygen uptake by the bacterial population in the natural water during t hours ($Bac\text{-}O_2$) can be calculated as follows:

When the size of the bacterial population is almost constant during the incubation time ($No = Nt$),

$$Bac\text{-}O_2 = \frac{No + Nt}{2} Qo_2 t$$

When the size of the bacterial population increases during the incubation time ($Nt > No$),

$$Bac\text{-}O_2 = \frac{Qo_2(Nt - No) \, t}{2.303 \, \log \frac{Nt}{No}}$$

where No and Nt are the initial and final numbers of bacteria in the natural (non-filtered) water, respectively.

Direct measurement of the respiratory activity of the plankton is difficult to apply in an oligotrophic water body. In this case the method originally reported for measuring total plankton respiration in the open sea by Pomeroy and Johannes (1966), which is described below, can be applied without any modification.

Plankton in a 20 l water sample is concentrated by filtering the water in which they are contained. For the filtration a membrane filter of $0.8\mu m$ porosity is used under a pressure of 5 cm of water. The filtration area is about 200 cm^2. As plankton is gently concentrated and the water is renewed continuously, depletion of oxygen generally does not take place and most of the plankton remain in the concentrated suspension. The surface of the filter is washed at the end of filtration with membrane-filtered lake water and plankton on the filter is resuspended by gentle brushing.

The plankton concentrate is placed immediately in a respirometer vessel and incubated in the dark at a temperature similar to the *in situ* temperature. When plankton is concentrated by using a membrane filter of smaller diameter, washing is not necessary. After filtration is completed the filter can be put directly into the respirometer vessel for determining the oxygen consumption by the oxygen electrode. When plankton concentrate is employed the oxygen consumption can be determined by the Winkler method in place of the electrode method.

3.3 Measurement of the decomposition rate by use of ^{14}C-glucose as substrate

For estimating the mineralization rate of organic matter in waters and bottom sediments the radioisotopic method is most sensitive. Although there are some limitations in the application of this method to natural environments, overall capacities of mineralizing activities can be estimated by this method.

A method (Kadota, Hata, and Miyoshi, 1966) using ^{14}C-glucose as substrate is outlined as follows: (1) Two hundred ml of a water sample or a water suspension of appropriately diluted mud sample (dilution is made with sterilized distilled water) is poured into a sterilized glass bottle painted with black enamel, and then 2—3 μ C of u^{14}C-glucose is added to the sample as a substrate; (2) after incubating for 2 hours at *in situ* temperature the bottle is connected with a distillation apparatus as shown in Fig. 7, and a small amount of H_2SO_4 is added to the reaction mixture in the bottle to gasify

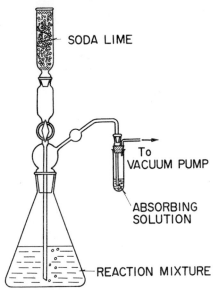

Figure 7. Apparatus for estimating the mineralization activity of lake water and sediment.

the $^{14}CO_2$ produced during incubation; (3) the $^{14}CO_2$ produced is completely expelled from the reaction mixture by aerating decarbonated air through the mixture, and is trapped by 7 ml of the absorbing solution composed of ethanolamine and ethylcellosolve (1:6); (4) radioactivity of this solution is counted by a liquid scintillation counter after the addition of 12 ml toluene, 0·08 g 2,5-diphenyloxazole (PPO), and 0·002 g 1,4-bis-2-(5-phenyloxazolyl)-benzene (POPOP).

Based on the counts thus obtained, the concentration of glucose in the mixture, and the amount of u-^{14}C-glucose added, the amount of glucose-^{14}C converted to CO_2-^{14}C by enzyme action can be calculated.

This value expresses the relative activity of glucose-oxidizing enzymes, mostly of microbial origin, in the sample.

It is generally thought that the glucose-oxidizing activities of water and mud samples give a good guide to their overall mineralization activities, because glucose is one of the most easily decomposable substrates for the majority of micro-organisms in aquatic environments.

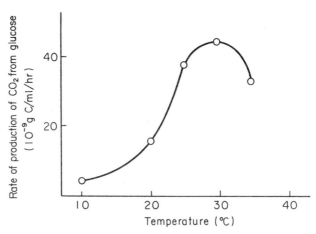

Figure 8. Effect of temperature on the mineralization rate of glucose in lake water.

As will be seen from Fig. 8, the mineralization rate was markedly influenced by the temperature. The effect of pH, however, was less marked than that of temperature.

3.4 Measurement of uptake of organic matter by micro-organisms

The application of kinetic parameters for the determination of microbial activities in water was first proposed by Wright and Hobbie (1965, 1966). By this method the uptake and metabolism of organic matter in lake water can be examined and the results give a measure of the rate of microbial decomposition of organic matter.

3.4.1 Principle of the method. Parsons and Strickland (1962) found that the uptake of labelled glucose and acetate in the sea follows the Michaelis-Menten kinetics. This uptake-process involves a substrate activation in the form of an enzyme-substrate-complex followed by the complete saturation of the enzyme. When all molecules of the enzyme are combined with substrate the maximal velocity of uptake is attained. A further increase in substrate concentration has no effect on the rate of uptake. The velocity of uptake (v) depends upon the kind of substrate, the enzyme and the affinity between enzyme and substrate. Where affinity is high, v is also high. The affinity between enzyme and substrate is characterized by the Michaelis-Menten constant (K_M), by definition the substrate concentration which gives rise to a half-saturation of the enzyme. At this concentration the velocity v is exactly one-half the maximal velocity V.

V is determined after the formula of Parsons and Strickland (1962).

$$V_t = \frac{c(S_n + A)}{C\mu\, t} \tag{1}$$

In this equation means V_t = velocity of uptake by a system (mg/1/hr), c = radioactivity of the filtered plankton (count/min), S_n = the concentration (mg/l) of the substrate present in the natural sample, A = concentration (mg/l) of added substrate (labelled and unlabelled), C = count/min from 1 μc ^{14}C in the counting assembly used, μ = number of microcuries added to the sample, t = time of incubation (hr). If the uptake of a substrate is mediated by a transport system the rate of uptake can be described in terms of Michaelis-Menten kinetics (Kepes, 1963), K_M is called here K_t, the transport constant. K_t is measure of the affinity of a substrate for an uptake system. The lower the K_t-values, the more effective is the uptake at low substrate concentrations. Thus the two parameters V_t (max. velocity of uptake, mg/1/hr) and K_t (transport constant, mg/l) represent important parameters of the microbial activity in water.

By transformation of equation 1 we get

$$\frac{C\mu t}{c} = \frac{S_n}{V_t} \quad \frac{(K_t + S_n)}{V} = T_t \tag{2}$$

T_t (in hours) is the time which is needed for the complete uptake of the substrate available in the water (turnover time). The uptake of substrate by algae seems to occur by another type of mechanism. Since the uptake of substrate does not show rate limitations with increasing substrate concentration as in Michaelis-Menten kinetics, thus the slope of the linear relationship indicates kinetics of a simple-diffusion (K_d instead of K_t) for the uptake of organic solutes by phytoplankton algae.

In studies on the uptake of ^{14}C-labelled glucose at various concentrations by natural microbial populations, the kinetic uptake pattern described above has not always been detected. It could be regularly found, however, when the water samples were pre-incubated with or without addition of unlabelled glucose (Vaccaro and Jannasch, 1967). This observation suggests that the predominance of one species or metabolic type in a given environment constitutes a prerequisite for obtaining uptake velocities that are amenable for kinetic analysis (suggested by the above authors).

3.4.2 Application of the method. To 50 ml samples in 100 ml glass stoppered bottles different amounts of ^{14}C-labelled glucose or acetate of known specific activity are added. As a rule, for the determination of the kinetic removal of the labelled substrates for each series four 50 ml samples are needed with addition of 50, 100, 200 and 400 μl labelled substrate (= 0·05, 0·1, 0·2 and 0·4 μC). Incubation time is adjusted so that not more than 5% of the substrate is removed in the most dilute sample.

Samples from eutrophic lakes need to be incubated at 15° C for 1 hour or at 3° C for 6—12 hours. The activity in these samples ranges from ca. 500 to 5000 count/ml. After incubation the samples are fixed with Lugol's acetic acid solution (by addition of four drops) and filtered using membrane filters (pore size 0·2μm). After filtration the filters are washed with distilled water (10 ml), airdried, desiccated and counted in a proportional counter of known counting efficiency. The number of bacteria and algae is determined by counting on membrane filters (cf. 4·2). Blanks for each concentration of labelled substrates are fixed immediately after adding the labelled compounds. The activity of the blanks is subtracted from that of incubated samples. Data from assayed filters are treated graphically and mathematically to obtain values for substrate affinity (K_t), maximum uptake velocity (V_t) and turnover time (T_t) according to the modification of the Michaelis-Menten enzyme kinetics by Lineweaver-Burk or Eadie (*e.g.* Fulton and Simmonds, 1958).

Applications of the methods of Wright and Hobbie to natural water systems are reported by several investigators (Wright and Hobbie (1965, 1966); Hobbie and Wright (1965 a, 1965 b); Allen (1967); Vaccaro and Jannasch (1967); Wetzel (1967)).

3.4.3 Estimation of natural substrate concentration. For a correct interpretation of the kinetic parameters knowledge of the natural substrate concentration is essential.

Estimation is possible by different methods:

1. By transformation of the Michaelis-Menten equation, the rapid graphical determination of ($K_t + S_n$) can be ascertained. After subtraction of K_t we get S_n (natural substrate concentration).

2. Bioassay with bacterial uptake kinetics to a filtered sample known quantities of an assay bacterium with a known K_t are added. After determination of $(K_t + S_N)$ (see above) the known K_t-value is subtracted from S_n (Hobbie and Wright, 1965 a).

3. To check the value of substrate concentrations obtained from kinetic data, the enzymatic determination of glucose by means of the optical test of Warburg is to be recommended (Bergmeyer, 1962). A modified method for the determination of various sugars in natural waters has been published by Ruchti and Kunkler (1966).

3.5 Calculation of the rate of microbial decomposition of organic matter from the heterotrophic uptake of $^{14}CO_2$

According to Romanenko (1965) heterotrophic bacteria, except for some spore formers, generally assimilate CO_2 at a ratio of about 7 μg CO_2-carbon per 1 mg O_2 consumed. About the same ratio is often found with the natural microbial population in aquatic environments. These observations indicate that the respiration rate of heterotrophic micro-organisms can be calculated from the rate of CO_2 assimilation. In the case of oligotrophic waters, in which the daily oxygen consumption is below the lower limit of the Winkler oxygen titration, this method is advantageous.

The principle of the procedure is as follows. The dark CO_2 assimilation value (A) measured by the radiocarbon method (cf. 5·3) is regarded as value of the heterotrophic CO_2 assimilation $(H, \mu g\ C/1)$. The rate of decomposition of organic matter by heterotrophic micro-organisms (D) can be calculated according to the equation:

$$D = 140\ H\ \mu g\ O_2/1 \text{ per day}$$

This method, however, cannot be applied to the waters which have direct contact with the anaerobic layer, where a large number of chemoautotrophic bacteria are living.

3.6 Decomposition and mineralization of nitrogenous organic matter

Proteins constitute the significant part of plant and animal cells and their decomposition and mineralization takes place mainly by microbial action.

Before this process can happen, the plant and animal cells must be made susceptible to attack from micro-organisms by preliminary autolysis. During autolysis most of the phosphates of the cells are mineralized, but only a small percentage of the nitrogen is liberated into solution (Golterman, 1960, 1964). The proteins are finally converted into CO_2 and NH_3 as the result of microbial action.

The mineralization of nitrogenous organic matter can be followed by measuring the production of NH_3 and CO_2 and the consumption of oxygen. This may be done either by geochemical methods or by the dark bottle method. The latter may be used only for relatively productive waters. Even here a long incubation time may be necessary but care should be taken that NH_3 is not converted into NO_3^- by too long an incubation time. A further disadvantage is that the amount of CO_2, NH_3 or O_2 taken up is always found as the difference of the value before and after the incubation time, which renders the method less precise.

Geochemical methods can be used for oligotrophic waters, especially when they are stratified. The difficulty here is the estimation of the amount of diffusion processes of NH_3, CO_2 and O_2 from the sediment into the water and from the water into the air and *vice versa*. The methods to be used may be found in the IBP Handbook No. 8. Suitable methods are:

For CO_2: 3.5.1 and 3.5.2 or for the most precise work, 3.4.2.
For O_2: 8.1.2 or 8.1.4 or for less precise work, 8.1.3.
For NH_3: 5.1.1.

3.7 Decomposition of organic matter in bottom sediments

3.7.1 Introduction. The remains of phytoplankton and the allochthonous organic matter sediment from the water column to the bottom of the water body. Here the sedimented organic matter is subjected to further decomposition. In lakes where dissolved oxygen is present in the bottom water decomposition in the upper layer of the sediment proceeds under aerobic conditions and results in the formation of carbon dioxide and water as final products. In deeper layers of the sediment, isolated from the oxygen penetration, anaerobic decomposition takes place. It results in the formation of carbon dioxide, molecular hydrogen and fatty acids as the main end products.

In other circumstances the anaerobic decomposition of organic matter in the sediment results in the formation of methane from fatty acid hydrogen and carbon dioxide. It is, therefore, possible to take the amount of the evolved CO_2 as a measure of the complete mineralization of organic matter if methane is counted as a product of the incomplete decomposition of organic matter. The amount of the consumed oxygen in an isolated volume of water in this case can be accepted as an indicator of the rate of aerobic mineralization of organic substances in the sediment. The difference between the total amount of CO_2 evolved as a result of aerobic and anaerobic processes and the amount formed during the aerobic process of decomposition can be taken as a measure of the rate of anaerobic decomposition.

The method described below was developed by Romanenko and Romanenko (1969), on the basis of the oxygen consumption methods reported by Hayes and Anthony (1959), and Gambarjan (1962).

3.7.2 Estimation of aerobic decomposition. The sample of sediment is taken with the aid of a sampler.* A column of sediment is taken from the sample with the aid of a glass tube 40 cm in length and 3·5 cm in diameter, using a weak vacuum in the upper part of the tube. This treatment does not disturb the vertical structure of the sediment. The bottom end of the tube is then closed with a rubber stopper. The tube with the sediment and the control tube are filled with water, which is taken from the bottom layer of water. It is important here to prevent the disturbance of the upper layer of the sediment. During the filling of both the experimental and control tubes the water is replaced several times. Then both tubes are stoppered with rubber stoppers so that no air bubbles escape.

The tubes are incubated in the dark *in situ* or at the simulated *in situ* temperature in the laboratory for 24 hours. At the end of the incubation period the tubes are held in the vertical position and are unstoppered. The water column over the sediment is mixed very carefully with a thin plastic rod having a loop at the end. This treatment eliminates errors connected with the uneven distribution of oxygen in the water column over the sediment surface. The mixing of water is made with both the experimental and control bottles. The water from the tubes is poured out with the aid of a tube into 60 ml bottles to determine the amount of oxygen in the water by the ordinary Winkler method. The volume of water in the tube is calculated geometrically.

* For a survey of suitable types of samplers see I B P Handbook No. 17: W. T. Edmondson & G. G. Winberg, A Manual on Methods for the Assessment of Secondary Productivity in Fresh Waters, 1971.

The oxygen consumption by the sediment (D_1) is calculated using the formula:

$$D_1 = \frac{n.H.l.1600.24}{t} \text{ mg } O_2/m^2/day$$

where: n = difference between the oxygen titration value of water of the experimental tube and that of control tube (50 ml water), H = normality of the thiosulphate solution, t = the duration of exposure in hours, l = the length in cm of water column in the tube. When the respiratory coefficient is accepted equal to 0·85, 1 mg of O_2 consumed corresponds to 1·59 mg of CO_2 or to 0·44 mg of the organic carbon oxidized.

3.7.3 Estimation of the rate of anaerobic decomposition. As was pointed out above, the anaerobic decomposition of organic matter in the bottom sediments can be assessed if the data on the amount of evolved CO_2 is known. Where O_2 in the bottom waters is absent and the process proceeds only anaerobically, all of the CO_2 formed has to be taken as a measure of anaerobic decomposition. The sampling and the arrangement of the experiments are identical to that described above. After the end of incubation the water is poured off from the tubes and the content of the free CO_2 is estimated in it by the titration with 0·01 N Na_2CO_3 solution in the presence of phenolphthalein as an indicator.

The calculation of the CO_2 formed (P) is made using the formula

$$P = \frac{n.H.l.1200.24}{t} \text{ mg } C/m^2$$

where: n = the difference between the titrations of 100 ml volumes of water from the experimental and the control tubes, in ml; H = normality of the Na_2CO_3 solution; l = length of the water column in cm; t = time of the incubation in hours. The value of the rate of anaerobic decomposition in the sediment (D_2) can be expressed by the following equation:

$$D_2 = P - (D_1.0\cdot44) \text{ mg } C/m^2/day$$

3.8 Measuring the dehydrogenase activity in bottom sediments by using triphenyltetrazolium chloride

Close relationships are generally found between trophic conditions in lakes and the dehydrogenase activity (DHA) of their sediments. The DHA-values are especially valuable for comparison with oxygen consumption as well as formation of carbon dioxide, redox-potential, evolution of methane or nitrogen and number of bacteria to be estimated separately in aliquot quantities of the same sediment.

Hydrogen from organic substances e.g succinic acid is transferred by enzymes to triphenyltetrazolium chloride (TTC) which is thus transformed to formazan. This red coloured compound is nearly insoluble in water but soluble in some organic solvents.

Reagents: (1) TTC-Reagent: Dissolve 0·9 g Triphenyltetrazolium chloride and 1·5 g Cadmium nitrate in 300 ml 0·1 M Tris-buffer (pH 7·5) containing in 360·9 ml 36·45 g Tris-oximethylaminomethane dissolved in 300 ml H_2O and 60·9 ml 0·1 N HCl. Finally the dissolved oxygen is eliminated by bubbling nitrogen through the reagent, (2) extraction mixture: 450 m lAcetone plus 300 ml Tetrachlorethylene.

Procedure: The mud cores are conveniently taken from water bodies using a sampler of the corer type.* Water covering the mud surface is sucked off and then the corer tube is opened along its length. The mud samples from different layers are taken by using a syringe (*S* in Fig. 9) the mouth of which fits to tube *b* of the incubation flask *F*. Meantime this flask is flushed with nitrogen gas to ensure that all the air is driven out. Just before fixing the syringe to tube *b* this must also be flushed with nitrogen. Instead of using the syringe mud samples can be filled into the flask by a spoon. However, in this case some oxidation of the mud occurs and produces inhibitory effects on anaerobic bacteria. After turning cock C_1 approximately 5 ml of the wet mud are pressed from the syringe into Flask *F*. Then cock C_1 has to be turned again in order to press out by nitrogen the mud remaining in tube *C*. Weighing of the mud sampler in the flask is generally unnecessary because the syringe delivers quite accurate quantities of mud. Again nitrogen is used for flushing out O_2 from tube *b*. Immediately afterwards the burette *B* filled with TTC-reagent is

---

* See previous footnote.

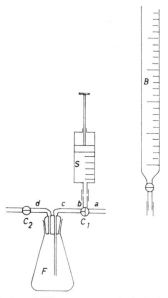

Figure 9. Incubation flask and filling apparatus inclusively for measuring dehydrogenase activity of mud samples.

fixed to tube b. Let 10 ml of the solution run into flask F and empty tube c by nitrogen flushing again. The cocks C_1 and C_2 are closed and the flask has to be incubated for 24 hrs on a shaking apparatus in the dark at 30°C. Open flask F and pour in approximately 40 ml of the extraction mixture, close again and shake for 1 hr on the shaking apparatus in darkness. Filter the mixture through paper (*e.g.* No. 589, 3 Schleicher and Schüll) into a measuring flask, wash the filter and fill up with extraction mixture under shaded light conditions and put in darkness immediately afterwards. A blank sample is run, *i.e.* by using the same procedure without TTC in the Tris-reagent. The extinction of the formazan solution is measured immediately at 540 nm in a spectrophotometer. The resulting values can be used for comparison by referring to the quantities of applied wet mud, the dry substance of which has to be determined separately. It is more advantageous to calibrate the extinction values in terms of H^+ by using 0·01 M ascorbic acid which delivers 0·01 equivalents H^+/L. The factor for computation may be *e.g.* 4·66 µg H^+ per extinction value 0·001 and has to be related to unit weight of wet or dry substance.

3.9 Measuring the evolution rate of gases from bottom sediments

The process of microbial fermentation in bottom sediments results in the production of methane and other gases. Their quantity depends on the concentration of organic substances and nutrients in the sediment. Remarkable differences are found in the gas-producing activity between muds from lakes of low and high productivity. The largest values, up to 1000 times those obtained with uncontaminated water bodies, occur in lakes receiving sewage waters.

a. Measurements by direct use of mud cores

Procedures: Take mud cores 20—50 cm in length and 10 cm in width together with the contact water layer 40—70 cm in height, by using a core sampler. The most convenient one for this purpose is the Iris-Mud-Sampler, the bottom of which is closed in the sediment by an iris-shutter (constr. Ohle, built by HYDRO—BIOS, Kiel, Fig. 10). Insert the special piston P

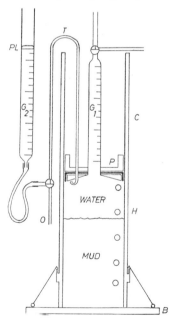

Figure 10. Experimental set with a core tube for measuring evolution rates of ferment gases.

which is combined with the gas burette G_1 and tube T as well. By this means a part of the lake water will be pressed out at the top of burette G_1 and through the connection by tube T with the levelling burette G_2 in which the surface of the water is covered by paraffinum liquidum. This surface has to be adjusted to the level of burette G_1 by opening the outflow O of tube T. Thus the piston P separates water and presses the sample of mud and water tightly. The height of the water column should be a few cm only. The core is incubated in the dark at 30°C or at such temperature as might be more representative of natural conditions. Then adjust the liquid level to point 'zero' in burette G_2 after adaptation to temperature of the whole apparatus. Read the volume of gas evolution in burette G_1 at short time intervals depending on the intensity of bacterial decomposition processes. Measure the total quantity of gas formation by reading the pressure in burette G_2 after having levelled it against the water surface in burette G_1. Compute the volume according to normal gas conditions (0°C, 760 Torr). The gas volume collected in burette G_1 is used for chromatographic analysis. The results are expressed as cm³ gas/m² per day or per month or other appropriate time interval. The volumetric composition of the gases (CH_4, N_2, CO_2, H_2, O_2) is also described.

Remarks: Additionally, the content and composition of the gases dissolved in the contact layer of water can be measured by gas-chromatographic methods. The water sample can be sucked out by inserting a syringe through one of the rubber stoppered holes (H) in the wall of tube S. The same holes are useful for injecting solution in experimental series.

b. *By use of homogenized mud samples in vitro*
This method* differs from the preceding one in homogenization of mud samples by mixing with bottom-near water during the incubation.

Procedure: The homogeneous sediment samples are placed in flasks of 250, 500, 1000 or 2000 cm³ volume depending on the expected volume of the gas evolved (Fig. 11), and water taken from the contact layer of the same lake is added. The quantity of wet mud applied is weighed in the flask. The volume of the mud should be at least ⅔ of that of the flask. A 100 cm³ burette working as a manometer is attached to the top of the flask. For incubation of the

---

* The procedure has been developed by A. Hamm, Munich, and is published in volume 19 of *Münchner Beiträge zur Abwasser-, Fischerei- und Flussbiologie,* 1971, edited by H. Liebmann, and published by R. Oldenbourg, München. The contributor (W. Ohle) thanks Dr. Hamm for allowing him to describe the method in this book.

Figure 11. Apparatus for determining activities of gas evolution of homogeneous mud samples.

mud the flask is placed in a water bath at 30°C in darkness. The contents are shaken and the mud whirled up occasionally or stirred permanently if possible. Gas evolved during incubation will drive water into the burette. The water level gives the quantity of gas at normal conditions after reduction to 0°C and 760 Torr according to barometric pressure, temperature and hydrostatic pressure of the water column in the burette. Results are expressed as cm^3 gas/g wet sediment, or dry substance, per time interval e.g. month or 100 days. The standard deviation of the values is 10% approximately.

The quantity of gases dissolved in the water cannot be determined precisely because the initial gas content of the water applied is unknown. However, for comparative work these errors can be neglected in the case of long time observations of a month or longer. According to the absorption coefficients at 30°C and 760 Torr of the participating gases, for equal volumes of gas and solution the volumetric errors would be 2·3% CH_4, 0·1% N_2 and 4·7% CO_2, assuming an average volumetric composition of 85·2% CH_4, 7·5% N_2 and 7·1% CO_2. The greatest error can occur with carbon dioxide, as

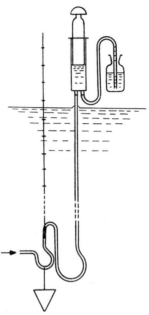

Figure 12. Apparatus for collecting water at various depths.

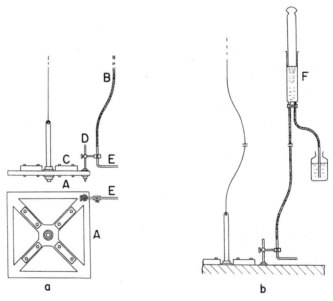

Figure 13. Apparatus for collecting water just above the bottom surface.

ought to be expected because of its high solubility in water. The errors will decrease with increasing size of mud sample. In both of the methods the adaptation time for microbial gas evolution is short if the samples are incubated at 30°C.

3.10 Geochemical methods for studies on the decomposition process of organic matter in natural waters

The process of decomposition of organic matter in natural waters can be analysed by determining the substrates products and intermediary metabolites of microbial decomposition of organic matter in water columns and in bottom sediments.

3.10.1 Mineralization rates of organic carbon and nitrogen

a. Sampling method

Apparatus: Water samples at various depths in lake water are collected by using an apparatus as shown in Fig. 12. Water samples just above the bottom surface are collected by using an apparatus which consists of a frame (A) and water inlet tube (E) connected with a long vinyl tube (B) and a pump (F) as shown in Fig. 13 (a) and (b).

Procedure: To collect a water sample at 10 cm above the bottom surface, for example, first adjust the mouth of the inlet tube at 10 cm above the frame surface and then lower the frame slowly with a rope to the bottom surface so as not to disturb the water layer. About 5 minutes later, collect the water sample by operating the pump as shown in Fig. 13 (b) (Koyama and Tomino, 1968).

b. Analytical methods of various inorganic elements in water

Dissolved oxygen and total amounts of other gases, except total carbon dioxide, are analysed by using the carbon dioxide method (Sugawara, 1939) modified by Koyama (Koyama and Tomino, 1968).

The apparatus consists of two main parts: A degassing vessel and a gas burette (Fig. 14). The degassing vessel (A) is a bottle holding about 250 ml. The gas burette (E) is a 10 mm tubing, graduated from 0 to 10 ml in steps of

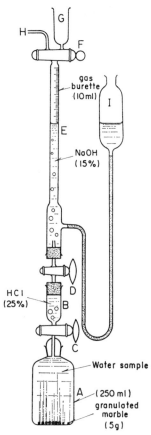

Figure 14. Apparatus for removing gases from water and for determining oxygen.

0·1 mm, and widened at its lower part into 15 mm bore, where a levelling bulb (I) is fitted.

Oxygen and total amounts of other gases are determined according to a procedure similar to that mentioned by Sugawara (1939). Nitrogen, argon and methane are analysed by gas-chromatography using oxygen as carrier gas.

The amount of denitrified (or fixed) nitrogen is calculated according to the following equation:

$$\Delta N_2 \text{ (ml/l)} = N_2(\text{obs.})(\text{ml/l}) - N_2(\text{theor.}) \quad (\text{ml/l})$$

Where ΔN_2 indicates the amount of denitrified (in case of positive value) or fixed (in case of negative value) nitrogen. N_2(obs.) indicates the amount of nitrogen dissolved in sample water. N_2(theor.) indicates the amount of nitrogen which is calculated from the determined amount of argon and the amounts of argon and nitrogen in water saturated with air, which were presented by Oana (1957). Argon is biologically inert and in water saturated with air, the concentration of argon and nitrogen are in constant ratio; nitrogen is subject to changes under biological influences. The amount of denitrified (or fixed) nitrogen calculated according to the procedure is subject to an unavoidable error of $\pm$ 0·1 ml/l.

Total carbon dioxide is analysed by using the hydrogen method (Koyama, 1953). The apparatus is shown in Fig. 15 (a) and (b). The experiment is carried out according to a procedure similar to that mentioned by Koyama (1953).

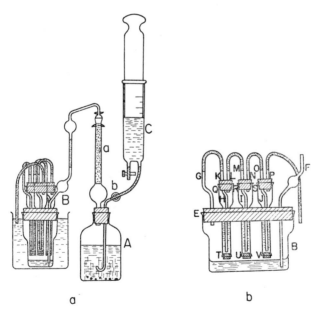

Figure 15. (a) Apparatus for determining total carbonic acid; (b) developed diagram of absorption vessel.

Ammonium is determined by the pyridine-pyrazolone method (Kruse and Mellon, 1953).

Hydroxylamine by the method of Endres and Kaufmann (1937).

Nitrite by the Griess' reagent.

Nitrate by the cadmium-copper reduction method (Wood, Armstrong and Richards, 1967).

Total manganese by using silver oxide as oxidizing reagent (Tanaka, 1951); Ferrous iron by the 2-2^1-bipyridyl method (Moss, 1942).

Organic carbon by a wet combustion method modified by Koyama (1954).

c. Estimation of mineralized carbon and nitrogen in a water column

Mineralized carbon (Miner. C) and mineralized nitrogen (Miner. N) at various depths are calculated according to the following equations:

$$\text{Miner. C (mg/l)} = [(\text{Total CO}_2 - \text{Orig. CO}_2\text{*}) + \text{CH}_4] \text{ (ml/l)} \times 0.536$$

$$\text{Miner. N (mg/l)} = \text{denitrified N}_2 \text{ (ml/l)} \times 1.24 + (\text{NH}_4^+ - \text{N} + \text{NH}_2\text{OH—N} + \text{NO}_2^-\text{—N} + \text{NO}_3^-\text{—N}) \, (\gamma/\text{l}) \times 10^{-3}$$

From the calculated values, the amounts of total mineralized carbon and nitrogen in a vertical water column are estimated.

3.10.2 Decomposition process of carbohydrate

a. The determination of dissolved carbohydrate

The water sample is filtered through a membrane filter (Millipore filter HA) and carbohydrate in the filtrate is determined by the phenol sulfuric acid method (Handa, 1966).

b. Determination of particulate carbohydrate

Particulate matter is collected on a glass fibre filter (pore size 1μm) through which the water sample is filtered. The material is treated with mineral acid to hydrolyse polysaccharides to monosaccharides and the total amount of the

---

* Orig. CO_2: Amount of total carbon dioxide which was not affected by the activity of organisms and thus existed originally in the water. The amount is assumed to be the total carbon dioxide in the water which was equilibrated with air by aeration for one hour at the prevailing air temperature.

monosaccharides is determined by the phenol sulfuric acid method (Hanada, 1966, 1967a).

c. Determination of monosaccharide composition of particulate carbohydrate
Particulate matter is collected on a glass fibre filter through which 100 liters of water sample are filtered. The material is treated with mineral acid to hydrolyse polysaccharides to monosaccharides and the monosaccharide composition is determined by paper chromatography (Handa, 1967b, 1969).

d. Solvent fractionation of particulate carbohydrate
Generally the solubility of carbohydrate in water and organic solvents such as ethanol and acetone primarily depends on the degree of polymerization of the carbohydrate. High molecular weight polysaccharides such as cellulose are insoluble in water, whereas low molecular weight polysaccharides such as algal starch and laminarin are soluble in water and are precipitated by addition of ethanol. Oligosaccharides are usually precipitated by addition of ethanol to the concentrated carbohydrate solution or by addition of ethanol and acetone to the dilute solution. Thus carbohydrates can be fractionated according to the following scheme.

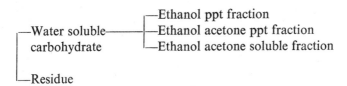

It will be assumed that the degree of polymerization of the carbohydrate fractions decreases in the following sequence: Residue > Ethanol ppt > Ethanol acetone ppt > Ethanol acetone soluble fraction.

3.10.3 Decomposition process of proteinous compounds

a. Determination of total proteinous amino acids dissolved in lake water

Sampling: Filter the water sample through a membrane filter (Millipore filter HA) immediately after collection. Keep the filtrate in a deep freeze for later use.

Reagents: (1) Dissolve 21 g of citric acid in 100 ml of 1 N NaOH and then dilute to 500 ml with distilled water. This constitutes a buffer solution of pH 5. (2) Dissolve 0·4 g of ninhydrin in 20 ml of methylcellosolve (ethylene glycol monomethyl ether). (3) Dissolve 0·04 g of stannous chloride in 20 ml of solution (1). (4) Mix 20 ml of solution (3) with 20 ml of solution (2). The mixture is prepared just before use.

Procedure: After melting the frozen samples at room temperature, evaporate 200 ml of the melted water sample to about 2—3 ml with rotary evaporator. Transfer the concentrated sample to a 40 ml test tube with 15 ml of 6 N HCl. After sealing the tube, hydrolyse the nitrogenous matter in the sample by heating at 100°C for 24 hours. After cooling, transfer the hydrolysed solution to a beaker and remove the greater part of the HCl by evaporation. Add 10 ml of distilled water to dissolve the residue. Neutralize the solution with NaOH solution to make hydroxides of Fe^{3+} and Al^{3+}. Filter off the hydroxides and evaporate the filtrate to about 30 ml. Introduce this solution to an anion exchange column (IRA 400, OH type and wet volume 20 ml) in order to separate amino acids from the solution containing NH_4^+ which also reacts with ninhydrin. Remove all of NH_4^+ from the column by passing 50 ml of distilled water through the column. Elute the amino acids in the column with 30 ml of 2 N HCl followed by 50 ml of distilled water (Moore and Stein, 1951). Evaporate all of the eluted solutions to dryness. Add 3 ml of distilled water to dissolve the residue. Neutralize the solution with NaOH solution and make the volume to 5 ml with distilled water, add 2 ml of ninhydrin mixture (4) and heat in boiling water for 30 minutes. During the heating period some amount of methylcellosolve is lost by evaporation. After cooling, make up the volume of the solution to 7 ml with 50% (v/v) of propyl alcohol. Measure the optical density of the solution at 570 nm.

b. Determination of different amino acids in proteinous matter dissolved in lake water

Sampling: Filter a water sample of 5 litres through a millipore filter (HA). Acidify the filtrate with concentrated hydrochloric acid to pH 1 and bring to the laboratory.

Reagents: (1) Solvent I: Methanol-water-pyridine (40:10:2). (2) Solvent II: n-butanol-methyl ethyl ketone-water-diethylamine (20:20:10:2). (3) Dissolve 0·5 g of ninhydrin in 100 ml of butanol saturated with water. (4) Mix 2·0 ml of a saturated solution of copper nitrate with 1·0 ml of 10% HNO_3 and make up to 100 ml with methyl alcohol.

Procedure: Evaporate 5 litres of the filtrate to 2—3 ml with a rotary evaporator. Hydrolyse the concentrated water sample with 6 N HCl in an ampoule and then remove HCl, Fe^{3+}, Al^{3+}, and NH_4^+ according to the procedure of item 3.10.3 a. Evaporate the residual solution to dryness and then dissolve the residue with 0·2 ml of water. Neutralize the solution with dilute NaOH (0·5 N) and make up to about 1 ml with water. Draw a base line along the length of filter paper (Toyo Roshi No. 51A or Whatman No. 51) (40 × 40 cm) 3 cm from the bottom. Draw another line 3 cm from the right edge of the paper. At the intersection of the base lines, apply two drops of the solution as a small spot. Staple the paper for ascending chromatography and place in a covered jar containing 150 ml of solvent I for 12 hours. After drying the paper, staple the paper for development in the second direction and place in a jar containing 150 ml of solvent II for 12 hours. After drying the chromatogram, spray it with ninhydrin solution (3) and heat at 80°C for 30 minutes to react amino acids with ninhydrin. Spray the coloured chromatogram with copper solution to make copper complex compounds (Bode, Hûbeer, Brückner, and Hoeres, 1952). Cut out the spots and extract each of the coloured copper complexes with a final volume (4 ml) of methanol. Measure the optical densities of the solutions at 510 nm. Calculate the amount of separated amino acids on a basis of their relative amounts and the total amount of amino acids.

c. Determination of different amino acids in particulate in lake or sea waters

Sampling: Collect particular matter on a glass fibre filter through which about 10 litres of lake water or 50 litres of sea water are filtered by the aid of suction. Keep the filter paper in a deep freeze for later use.

Procedure: Cut the filter paper into small pieces and transfer the pieces to a 40 ml test tube with 15 ml of 6 N HCl. After sealing the tube, hydrolyse the nitrogenous matter and then remove HCl, Fe^{3+}, Al^{3+} and NH_4^+ according to item 3.10.3 a. Determine various amino acids in the hydrolysate according to item 3.10.3 b.

4
The Determination of Microbial Numbers and Biomass

The microbial biomass in natural waters has largely been estimated by various direct and indirect techniques of determining numbers and sizes of bacterial cells. A quite different approach involves determination of ATP. The present chapter deals with some of these methods and also with techniques of microbiological sampling.

4.1 Sampling techniques

For the determination of total microbial numbers by direct microscopic methods and for the measurements of biomass of the living micro-organisms by the ATP-method the water samples can be collected with a clear plastic water bottle usually used for limnological work. Sometimes it is recommended that the bottles be rinsed with 1:4 HCl before use. In case of a viable count sterilized microbiological samplers should be used.

In contrast to ordinary water samplers for chemical studies, microbiological samplers have to be sterilizable. The sampling gear must permit lowering of the collecting vessel to the desired depth, opening, filling and closing it by a triggering mechanism. Different types of microbiological sampling vessels are: (a) evacuated bottles, (b) rubber bulbs, (c) plastic bags, and (d) syringes. The following sampling devices have been found practical for freshwater and marine studies.

a. Bottle sampler
This simple apparatus (Fig. 16) can be used for microbiological sampling to depth of 50 to 60 meters, Sorokin (1960). An evacuated sterile bottle (Fig. 16) is fastened to a heavy frame (1) by a clamp (2). A curved glass tube (4) penetrates a rubber stopper and is connected to a thick-walled rubber tube (5) of 10 to 15 cm in length which is closed by a sealed glass capillary (9). Another clamp (6) holds the capillary to a plate (7). When the capillary is broken by a

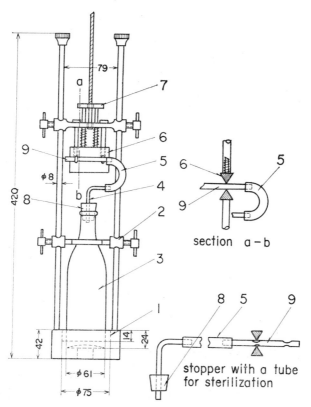

Figure 16. Bottle sampler for microbiological work (explanation cf. text).

messenger, the sample is sucked into the evacuated bottle through the open end of the rubber tube that flips some distance away from the sampling gear.

Another type of sampling vessel (Sorokin, 1964) consists of a glass ampoule that facilitates taking series of samples at the same time (Fig. 17). As the messenger breaks the capillary and the sample is taken, another messenger is released in order to trigger the next sampler. In field studies, the cylinder can simply be steamed for sterilization. After sealing of the two capillaries, the necessary vacuum forms by condensation.

b. Rubber bulbs

Rubber bulbs of ca. 300 ml volume have been introduced as sampling vessels by ZoBell (1946) and Sieburth (1963). They can be autoclaved and evacuated

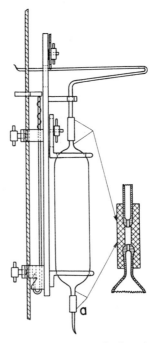

Figure 17. Glass ampoule water sampler for microbiological work.

simply by deflation. Arrangement of the glass capillary and fastening to the cable is similar as shown in Fig. 17. These rubber bulbs can be used to a depth of several hundred meters.

c. Plastic bags

If a non-elastic plastic material is used for a sampling vessel, filling must be operated by an external source of power. This is realized in the Niskin sampler (Niskin, 1962). Upon triggering by a messenger, two plates open in V-fashion by spring power thus inflating a plastic bag. The intake tube of silicone rubber is cut by a sharp blade and, after filling of the bag, is closed by a clamp. There are 3 litre and 5 litre models of the sampler. The temperature can also be taken at the time and place of sampling.

d. Syringe-type sampler

In the samplers described before, contamination of the water sample may result from breaking the glass capillary or from cutting the rubber tube intake by a non-sterile object. Furthermore, the samples are taken near the sampling gear and the hydro-wire might have been heavily contaminated by passing through the neuston layer. It has been shown that *Escherichia coli*, contaminating the surrounding surface water of a research vessel, was carried down to great depths by the sampling gear (Jannasch, 1968).

To prevent such contamination, a special sampling device has been constructed and tested employing *Serratia marinorubra* as a tracer organism (Jannasch and Maddux, 1967). The sampler consists of a plastic syringe which is operated by gravity. The sample intake is a silicone rubber tube of 10 cm in length which is covered by a piece of dialysis tubing before autoclaving. This cover is removed at the site of sampling exposing the sterile intake tube just before the sample is taken. By means of a vane, the sampler rotates freely around the hydro-wire on a swivel, always heading into the current. When the sampler is triggered by a messenger, (a) the intake tube swings about 60 cm away from the hydrowire, (b) the intake tube is uncovered, (c) the plunger sucks the sample into the syringe, and (d) the intake tube is closed.

Experiments showed that heavy artificial contamination of the sampling gear with the tracer organism did not result in a contamination of the sample. The sampling vessel contains no air and can be operated at any depth. The disadvantage of the present sampler, which can easily be reproduced in a workshop, is the small size of the sampling syringe (50 ml). There is no problem, however, in constructing a sampler using a larger syringe on the same principle.

e. Sampling of sediments

There is no completely sterile sediment sampler. Cores are taken by hand or at the end of a hydro-wire (Phleger corer), whereby micro-organisms from the surface of the sediment can be carried down on the outside of the core. After retrieval, however, the core can be frozen in the liner in order to prevent any further contamination of the inner part. At thawing, samples can be taken from various parts of the core with sterile objects for further studies.

With the aid of a sterile tube (Fig. 18) the subsamples of sediment for the microbiological studies can be taken from the ordinary grab sample. Sucking of the subsample by use of a rubber tube 1 into the sterilized glass tube 2 has to be done simultaneously with its insertion into the sediment so that the levels of the sediment in the tube would be the same as that in the grab. In this case stratification of the sediment in the tube is not disturbed. Then the

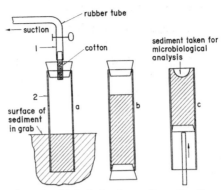

Figure 18. Device for microbiological subsampling of sediments from the grab sample: (a) Scheme for subsampling; (b) a tube containing subsample; (c) sampling for microbiological analysis from subsample.

tube is drawn out and both ends are stoppered with sterilized rubber stoppers. These subsamples can be frozen and stored until use. A portion of the sediment for the analysis is taken by gradual pushing out (Fig. 18, c).

4.2 Direct microscopic counting of micro-organisms

The direct microscopic count is an important method for the quantitative study of natural microbial population because this method gives the number of all the cells of micro-organisms present in waters or sediments (Razumov, 1947; Kuznetsov, 1959). The microbial population of a water is counted on the membrane filters after filtration of a definite volume of water sample. The number of micro-organisms in bottom sediments can be counted on slides (Winogradsky method) and also on membrane filters.

When differential staining is not attempted the water samples can be fixed with formaldehyde (2% final concentration) for storage. The pore size of the filters to be used for the direct microscopy must not exceed 0·3 µm. The working surface of the filters must be smooth and free of contamination. A Czechoslavakian filter VUFS (pore size: 0·1—0·3 µm) is recommended for this purpose. In cases where a clean filter, free of contamination, is not available, it is necessary to determine a correction factor for the filter to be used.

Before use the filters are numbered with a pencil and boiled in distilled water. The filtration is made with the aid of a glass funnel of a type of the

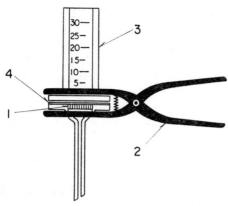

Figure 19. Glass filtration funnel of the 'Millipore' type: 1—porous glass plate; 2—metallic clip; 3—glass calibrated funnel; 4—millipore filter.

American 'Millipore' funnel (Fig. 19). For the estimation of the correction factor for a contaminated filter a round piece of plastic film of 0·1—0·2 mm thickness with a square opening of 1—2 cm² in the centre is placed under the filter. The dimension of the opening depends on the working surface of the funnel used. Then the microscopic examination is made with the 'working' surface of the filter and with the surface covered with the film. The latter is made for the estimation of the error due to the previous contamination on the filter surface. The volume of water filtered through the filter is dependent upon the trophic level of the water body to be examined. For oligotrophic waters 10—25 ml of water is passed through 1·5—2 cm² working surface of the filter; for mesotrophic waters the volume is 5—10 ml, for eutrophic ones the volume should be 2—5 ml. For heavily contaminated waters a volume of 0·5—1 ml is appropriate. When a small volume of water is employed 2—3 ml of previously filtered water is added to the funnel to provide a uniform distribution of micro-organisms on the filter surface.

The micro-organisms on the filters are fixed in formaldehyde vapour by placing the filters in Petri dishes with several drops of concentrated formaldehyde on the upper lid, or by heating (20 min, 70° C). The filters are then dried and stored until the counting.

Before examination the micro-organisms on the filters have to be stained. A sector of the filter is cut, and placed for staining into a Petri dish at the bottom of which is placed a round piece of 2—3 layers of plankton net or filter paper moistened with 3% erythrosin solution in 5% phenol. The erythrosin solution can also be prepared by dissolving 0·5 g of the dye in 96% ethyl alcohol. For the staining with phenol-erythrosin stain 0·5, 1 day at room temperature, and 20 min at 65°C in the moist atmosphere of a desiccator is required. With the use of alcohol-erythrosin it takes 1—2 min. The stained filters are then decolourized by placing them on the surface of the plankton net fixed to the surface of water in the cup or on the moistened filter paper. The filters are decolourized to a weak pink colour. The erythrosin staining can be contrasted by subsequent staining with the diluted carbolic fuchsin ($\frac{1}{300}$). The diluted fuchsin is heated at 60°C. The filter, previously decolourized as described before and dried, is placed on the surface for 20—30 s. Then the filters are dried again.

For the microscopic examination the filters are placed on a drop of immersion oil placed on the surface of a slide. Onto the surface of the filter a further drop of the oil is put. By this treatment the filter becomes quite transparent. The filter is covered then with a cover glass. Microscopic examination of the preparations is usually made with an immersion objective × 90 and ocular × 15 in the binocular microscope under green light. Inside of one of the oculars a grid is placed. For this purpose it is convenient to use a grid which has 25 sections in an area 25—35 mm². The number of bacteria (N) in the initial sample can be calculated using the formula

$$N = \frac{S \cdot n \cdot 10^6}{s \cdot v} \text{ cells/ml}$$

where: S = area of the 'working' surface on the filter (mm²), s = area of 1 cell in the ocular counting net, measured with the objective micrometer under the same magnification of the microscope (μ^2), n = average number of bacteria per 1 section of the net, v = volume of water filtered.

In fresh water samples 9 sections should be counted of the total 25 sections of the counting net placed by cross diagonal directions. If the number of bacteria on the filter is less than 5 per 1 section of the net it is necessary to count more fields. To determine the number of fields to be counted statistical methods can be used.

The statistical homogeneity of distribution of cells on the filter can be checked by the chi square test. Since, however, the bacterial cells are often arranged in pairs, short chains or clumps, a slight deviation from the stochastic distribution must be considered. Hence, if the calculated value of chi square does not exceed $P = 0.01$ for the appropriate degrees of freedom, the filtration technique can be considered adequate. This chi square must be calculated at least for ten microscopic fields chosen at random in different parts of the filter. Where bacterial clumps are large and/or frequent, single cells must be counted apart from the cells in clumps and the chi square calculated on the separate figures.

The precision of direct microscopic counts increases with an increasing number of cells, because the deviation is equal to the total number of the cells counted. The confidence intervals can be calculated by means of the graphs by Cassell (1965), provided that the distribution of cells on the filter is homogeneous (Fig. 20).

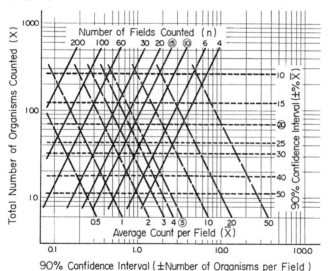

Figure 20. Schedule of precision at the 90% level of confidence.

4.3 Direct count of micro-organisms in bottom sediments

The Winogradsky method used for soil microflora was modified by Kuznetsov and Romanenko (1963). A sample of the sediment (0·5 ml) is taken with a sterile glass tube and transferred to a flask with 50–100 ml of 0·005 N KOH. For lime containing sediments the sample is treated with 2% HCl and then neutralized. The flask with the suspension of the sample is shaken for several minutes and then is left for 0·5—1 min to settle the big particles. A 0·1—0·2 ml portion of the suspension is placed on the surface of a microscopic slide which has been thoroughly cleaned with alcohol and ether. On it a drop of 0·05% agar solution previously filtered through the millipore filter is added. The mixture is carefully and uniformly spread over the slide surface on the area 6 cm². The preparation is dried, fixed by flaming or by absolute alcohol and stained. The carbolic erythrosin solution is poured onto the slide preparation and then heated until there is slight vaporization. Then the dye is washed off and the diluted carbolic fuchsin solution (1/300) is poured on it. After heating again the preparation is washed and dried. The microscopic examination is made under the oil immersion. The counting of bacteria is made with the ocular grid.

The number of bacteria in the suspension of sediments prepared as described above can be counted also on membrane filters. After shaking the suspension is stood for 3—4 min to settle large particles. A quantity of 0·1—0·2 ml of it is poured into the filtration funnel together with 1—2 ml of the previously filtered water. The subsequent analysis is as described before (4.2).

4.4 Determination of the cell size of micro-organisms for the calculation of biomass

At the same time as the total count of micro-organisms is estimated on membrane filters, the ratio of different morphological types of cells can be determined in the following simple manner: (a) long rods, (b) short rods, (c) coccoid forms, and (d) yeast-like cells. The two dimensions of each cell type should be measured microscopically for calculation of the volume. The mean volume should be computed on the basis of individual volumes, not from the mean length and width. When a projection microscope is used, cell lengths of about $0.25\mu m$ or a volume of about $0.015\mu m^3$ can still be determined with a fair degree of accuracy. Since the volumes of microbial cells in natural waters have been found to be in the range of 0·05 to $2.0\mu m^3$, the precision of the estimates will not exceed 8—32% of the mean.

Stained preparation on glass slides can also be used. Cells may be concentrated in the water samples by partial filtration on membrane filters or by centrifugation. The filtration in this case has to be stopped just before the moment when 0·5—1 ml of water remains in the funnel. Fixing and staining may decrease the original size of the cells by a factor of 2·5 (Troitsky and Sorokin, 1967). The average values of cell volumes have been shown to be 0·3—0·6μm in oligotrophic, 0·5—1·2μm in mesotrophic, and 1·0—2·0μm in eutrophic waters.

Using concentrated suspension of cells, volumes of living cells can be measured by a phase contrast microscope. The size of cells can also be measured with magnified microphotographs.

The variation of cell size of bacteria in one individual sample (Table 1) is rather higher than the variation of the mean cell volume during a whole year (Table 2). The size determination of bacteria in 5 samples during one year

Table 1. The variation between the sizes of bacteria in one sample (after V. Straskrabova)

Locality	Month	Coefficient of variation*	95% confidence limits	
			50 measurements	100 measurements
Slapy reservoir	January	81·0	32·3	16·2
,,	May	98·8	39·7	18·8
Klicava reservoir	February	90·5	35·2	18·2
,,	March	132·3	51·2	25·7
Berounka river	May	91·4	36·7	18·5
,,	November	108·5	44·2	20·7
,,	December	94·8	37·5	19·5
Treated sewage	October	113·2	44·6	21·8

* Standard deviation expressed as percentage of the mean, 95% confidence limits expressed as percentage of the mean.

Table 2. The variation between the mean volumes of bacteria during a year* (after V. Straskrabova).

Locality	Year	Coefficient of variation**	95% confidence limits	
			1 sample	5 samples
Slapy reservoir	1966	20·0	80·8	16·2
,,	1967	25·2	100·7	20·2
Klicava reservoir	1966	21·8	88·3	17·6
,,	1967	18·3	70·9	14·2

* 50 measurements of each sample.
** See Table 1.

(100 measurements in each sample), reached a 95% confidence limit in the range $\pm$ 20—30% of the mean volume. To obtain the raw biomass of the microbial population (B) the total number of cells counted by the direct microscopic method (N, 10^9/l) is multiplied by the mean volume (V, μm^3: B = V.N mg/l. To calculate the dry weight and the weight in terms of carbon the coefficients 0·2 and 0·1 (Troitsky and Sorokin, 1967) can be used, respectively.

4.5 Direct count using fluorescent microscopy

This requires a fresh sample or one preserved with 5% glutaraldehyde and kept in the cold and dark. Samples preserved thus can be studied up to a week after they are taken. The sample is concentrated if necessary, preferably by continuous centrifugation (Kimball and Wood, 1964), or may be filtered onto a glass filter (Whatman, GFC) in which case it must be studied by reflected light through a well-designed system.

A suitable aliquot is taken—determined by previous experience and put in a chamber with a thin glass base, and of known volume *e.g.* Petroff-Hausser bacterial counting chamber. Either the original sample or the aliquot is stained with 1/20,000 to 1/50,000 acridine orange on the alkaline side of neutrality. The living bacteria should fluoresce green and dead bacteria and organic detritus should be brick red. If photosynthetic organisms are present, the chloroplasts should be bright scarlet (different from dead material) and the nucleus and protoplasm green. If the preparation is poor, it is probable that the stain is too strong.

The slide is then counted in light with a wave-length around 450 nm (not ultraviolet, which spoils the picture). It is possible to estimate the number of bacteria in the original sample if the area and depth of the slide are known. As the lighting is critical, oil should be placed between the condenser and the bottom of the slide.

Fluorescence counting is possible with a monocular microscope with an Abbe condenser and a lamp with a flat tungsten filament such as the Unitron high intensity lamp. Better resolution is obtained, however, with a mercury arc lamp and a quartz condenser. With this arrangement, inclined binocular microscopes can be used. The best filters so far tested are the 'Fluoreszenz B' series (BG12 and OG1) offered by the Wild Company of Switzerland.

The fluorescence method is useful for counting bacteria and algae in eutrophic water, in sediments, and in grazing studies, especially of copepods, protozoa, and nematodes. It is necessary to allow the eye to become dark-adapted before beginning to count.

Sediments may be studied by making suitable dilutions in a fine capillary, by making a series of equidistant marks on the tube with an india ink pen, drawing up one aliquot of the sediment with the required number of aliquots of water. Or a small amount of sediment may be diluted with a known volume of water and then sucked up into the capillary. The sample is then put into a suitable cell, covered with a slip and examined under the microscope. If depth and diameter of the cell is known, the bacterial counts can be made quantitative.

Other available methods which have so far not been adequately tested are the fluorescence antibody method, the direct reflected light method for sediment surfaces, and the autoradiograph method.

4.6 Use of the electron microscope

The use of the electron microscope in the study of the microflora in lakes revealed new kinds of micro-organisms which had not been detected before (Nikitin and Kuznetsov, 1967). For the studies of aquatic microflora, magnifications of 10,000—50,000 are generally used.

The water samples to be examined by electron microscopy are made clean by filtration through membrane filters (pore size, 4—6μm). Micro-organisms in the filtrate are concentrated by filtration using finer membrane filters (pore size 0·1—0·2μm) or by use of a centrifuge (15 × 10^3 rpm for 30 min).

At the bottom of a specially designed Petri dish with a glass tube (Fig. 21) a series of copper net planchets are placed. These planchets are previously cleaned by boiling in 10% NH_4OH or by rinsing with conc. HCl and washed with water. The dish with the planchets is then filled with distilled water. On the surface of the water a drop of 1% collodium solution in amylacetate is placed. After the formation of a fine film on the water surface just above the planchets, drops of a concentrated cell suspension of micro-organisms are placed. Then the dish is incubated at 28°C for 17—24 hours to remove mineral salts from the suspension by dialysis. The micro-organisms are fixed during the incubation with formaldehyde vapour. A strip of paper moistened with formaldehyde is placed on the inner surface of the Petri dish. After the

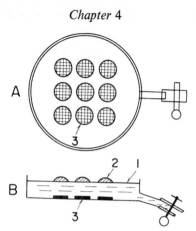

Figure 21. Preparations on the net planchets for electron microscopy. A—view from above; B—view from a side; 1—collodium film; 2—cell suspension of microorganisms; 3—net planchet.

dialysis, water in the dish is poured off through the side tube. By this treatment drops of the suspension are settled on the corresponding planchets. The drops on the planchets are then dried and subjected to further examinations.

The planchets can also be prepared separately. On the surface of the planchets covered with collodium film, drops of cell suspensions of microorganisms are placed. The suspensions are previously dialysed on the film placed on water surface and then transferred to prepared planchets with the aid of a fine pipette.

The dried preparations are then subjected to the metal shadowing. This is achieved by exposing the preparation to a stream of vapour of metals such as Cr, Pt or Au (Nikitin *et al.*, 1966).

Contrast can also be obtained by treatment of the preparation with 1% w/v solution of phosphorus wolframate (FWA) at pH 7.2. After a few seconds the solution is removed by filtration and the preparation is dried. 'Positive' staining is achieved by incubating the preparation with 1% FWA (pH 1·7) for 2—4 s. The FWA combines with proteins of the cells. The FWA solution after staining is removed by placing the planchet on the surface of a filter paper. The same effect can be obtained by use of a neutral solution of FWA, if the staining is made for 5—10 s and then the preparation is carefully washed. In this case the background film is clear and the microbial cells are contrasted by FWA. In place of the FWA, uranyl acetate, osmic acid, or lead salts can be used for this purpose. For the preparation of micro-organisms living in waters of high salinity an additional treatment is required, because some halophylic micro-organisms living in such waters are destroyed during the dialysis. In such cases the 'heat attachment' of microbial cells to the collodium film (Stefanov, 1962) can be used. A drop of water without any previous treatment is placed on the surface of aluminium foil with rolled edges. On the surface of the drop the planchet covered with collodium film is placed. The foil is then placed on the surface of hot water (50—55°C). Some of the micro-organisms fasten to the surface of the film. After 2—4 min the planchet is removed from the drop. The planchet is dried by removing water with filter paper. Then the planchet is contrasted by staining with 1% FWA (pH 1·7) for 1—2 s.

In the case of sediment the sample is suspended in 10 volumes of 0·0004N NaOH and then a portion of supernatant of the suspension is used for making the preparation for electron microscopy (Nikitin, 1964).

4.7 Capillary method for the study of periphytonic organisms

The periphytonic micro-organisms grow on the surface of submerged objects as well as on the particles of bottom sediments (Perfiliev and Gabe, 1969). Some of these organisms often move from the particle surface into the water and *vice versa* when environmental conditions change.

If the ordinary submerged slide method is used, some of the periphytonic micro-organisms could be washed out from the slides and the periphytonic community could be disturbed when the slides are pulled out of the water. The capillary periphytonometer and the capillary peloscope are used for the study of periphytonic microflora in water and that in sediments, respectively.

4.7.1 Capillary periphytonometer by Perfiliev. The periphytonometer consists of two parts, *viz.* the glass plates with capillary trapping equipment and the metal base. The shape and size of the glass plates are similar to those of ordinary microscopic slides. A piece of glass band (64·0 × 2·0 × 0·5) is used as a substrate for providing solid surface to periphytonic micro-organisms. The band is preserved by a capillary (rectangular with trapeziform section), which slides freely on the glass band, but does not touch its upper surface. The free slot space between the internal upper wall of the capillary and the upper surface of the band ensures the undisturbed state of the periphyton.

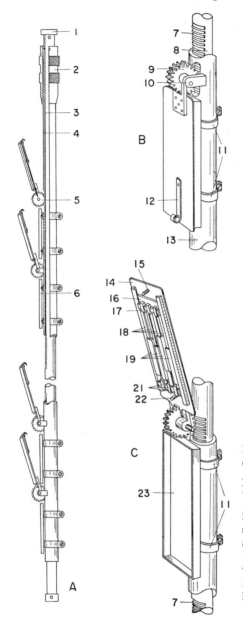

Figure 22. Periphytonometer with cremailliere closing equipment: A—apparatus with opened casettes; B—closed casette; C—opened casette; 1, 7, 8—internal tube with cremailliere; 2, 4, 13—external tube with cuts and screw mechanism, which regulate the angle of inclination of the casettes; 6, 11, 23—preserving leaves of casettes, which are fixed on the external tube; 3, 5, 9, 10, 12, 14, 15, 20, 22—rotatory leaves of the casettes with cogwheel; 16, 19, 21—trapping plate.

The capillary is 30 mm long, and has a thick bottom and very thin upper walls. The periphyton samples thus prepared are examined under the microscope. The bands with preserving capillaries are fastened on the glass plate by means of glass planks in such a way that the capillaries cannot touch each other (Fig. 22: 16—21). When the trapping plates are opened, the preserving capillaries are on the lower end of the bands and the upper parts are opened. After exposure, the trapping plates are turned by 180° and the preserving capillaries close the exposed parts of the bands.

The metal part of the periphytonometer consists of a supporting dualuminium (or better non-corrosive steel) tube with the casettes for trapping plates. Each casette consists of two leaves, the rotary leaf with the slots for glass plates and the stationary leaf, which is fixed on the supporting tube and looks like a flat box (Fig. 22: 23). Glass plates are fastened on the rotary leaves by means of springs. The casettes must be closed when the apparatus is lowered into the water. For the time of exposure the rotary leaves of the apparatus are raised up and the trapping plates opened.

There are 10 rotary casettes disposed along the supporting tube. The internal tube is moving inside the supporting tube, which has a clearance of 0·5 mm. Both tubes have the same length of 1·4 m. The wall of the external tube has longitudinal cuts at the points of fastening of the rotary casettes. The wall of the internal tube has transversal incisions. These parts of the tube can be used as the teeth of a cremailler. The gearwheel on the axle of the rotary leaf of each casette is connected with transversal cuts of the internal tube.

Downward movement of the internal tube opens all the casettes and prepares the apparatus for the exposure. The opposite movement of the internal tube closes the casettes. The movement of the internal and external tubes is regulated by two ropes. The inclination of plates during exposure is regulated by a screw mechanism, which is disposed on the upper end of the supporting tube.

After exposure, the periphytonic microflora can be investigated in the living state and after fixation. The space between the upper surface of the band and the upper wall of the capillary is filled with Canadian balsam and the bands fixed on the surface of microscopic slide.

A simpler model of the periphytonometer may be used in shore conditions. The rotary leaves of the casettes are fixed on the tube by means of a yoke. A load is fixed on the lower end of the supporting tube and synchronous opening and closing of the casettes can be produced by pulling the two opposite ends of the rope (Fig. 23). There is a side cup on the lower end of the supporting

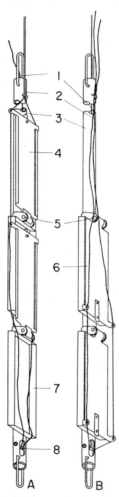

Figure 23. Simple model of periphytonometer: A—casettes are opened; B—casettes are closed; 1—end of rope by means of which the casettes may be closed (by tension of the rope); 2—end of rope by means of which the casettes may be opened; 3—tube with fixed casettes; 4—rotary leaf of casette is opened; 5—yoke with the axle of rotatory leaf of casette; 6—rotatory leaf of the casette is closed; 7—preserving leaves; 8—roller of the rope; 9—loop for a load.

tube. The rope passes through the side cup of the tube and over the block. When the periphytonometer is submerged in the water, the lower end of the rope is held. Opening of the casettes can be produced by pulling the upper end of the rope, which is connected with the leaf of the upper casette.

4.7.2 Capillary peloscope of Perfiliev. In bottom sediments micro-organisms often form microzones. Some of them are very thin—several micrones. The microzonal distribution of micro-organisms, when disturbed by mechanical factors, can be restored rather quickly. The apparatus devised by Perfiliev allows examination of the microbial layers of bottom sediments in an undisturbed state.

The peloscope consists of a set of 4—5 flat glass capillary cells. Each cell is composed of five rectangular canals arranged in one row (Fig. 24.2). The cells are attached to a special glass holder with a millimeter scale. The sample of bottom sediment must be thoroughly mixed, put into a glass vessel and filled with water of the same source. The vessel is covered with glass and held in the dark. When the sedimentation of mud particles is completed and the water becomes transparent the peloscopes can be put into the sediments. The capillary canals must be arranged vertically and must not contain bubbles of air. The peloscope can be put into the sediments at the desired depth by means of a millimeter scale of the holder. The lower end of the holder must be set against the bottom of the vessel (Fig. 24:1). After removal from the sediment the peloscopic cells can be examined and then returned to the same place.

In 1—2 weeks after mixing, the primary homogeneous sample usually divides into microlayers and the formation of microzones begins. The microbial layers, formed inside the capillaries, can be studied under the microscope without disturbance. The microbial population inside the peloscopes is protected from the masking effect of sediment particles and from mechanical damage. The exposure time of peloscopes may be different (from several days to several months) and depends on the duration of the process of forming microzones, Gabe *et al.* 1967.

The investigation of the periphytonic microflora in a living state has to be done immediately after taking up the peloscopes from the sediment, because changes in the physico-chemical conditions inside the capillaries may influence the character of the distribution of micro-organisms in the peloscopes.

The peloscope capillaries have to be carefully washed and freed from the rubber ring. Each cell is wiped dry by soft cloth and then is placed on the surface of a glass slide for the microscopic examination (Fig. 25).

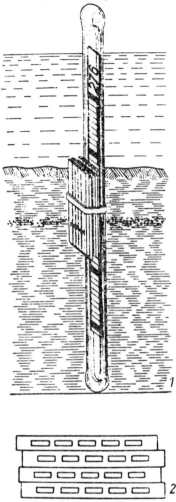

Figure 24. Capillary peloscope: 1—vessel with peloscopic cells, fixed on the holder submerged into mud; 2—cross-section of the capillary cells.

Figure 25. Capillary cells on the slide.

4.8 Enumeration of viable cells of micro-organisms by plate count technique

4.8.1 The general procedure. Using the plate count method the number of micro-organisms in a sample is derived from the number of colonies which ultimately grow on a plate. This method has many limitations and essentially gives a relative number of limited groups of heterotrophic micro-organisms. Yet, if its limitations are appreciated the technique can yield useful information.

A typical practical procedure would be as follows: (1) Make a known single dilution or several known serial dilutions of the sample with a sterile diluent fluid. By this procedure progressively less dense suspensions of micro-organisms are prepared. Certain water samples may not require dilution; (2) take a known volume of selected dilutions and ensure that the micro-organisms therein are brought into intimate contact with a sterile nutritious agar medium. Commonly used procedures are (a) the 'surface drop' technique (Miles and Misra, 1938; Badger and Pankhurst, 1960). Drops of suspension, delivered from a calibrated pipette, are deposited on the surface of pre-dried agar medium; 10—12 drops can usually be accommodated on the surface of the agar in a 10 cm Petri plate, (b) the 'spread-plate' technique. A small volume (0·1—0·2 ml for a 10 cm Petri plate) is spread over the surface of the pre-dried agar by means of a sterile glass spreader. For both these methods it is essential that the agar surface be carefully dried to ensure that the microbial suspension soaks rapidly and completely into the agar. The plates should be held at 5°—10°C during the absorption period to curtail multiplication in the surface film. If a surface film is still present during incubation the system resembles a liquid culture on a solid surface. The outcome is usually confluent growth over the agar surface instead of discrete colonies of cells, (c) the 'poured-plate' technique. A volume of suspension (usually not less than 1 ml for a 10 cm Petri dish) is transferred to a sterile Petri dish. A volume of sterile, cooled but molten agar (usually 10 ml for a 10 cm Petri dish to give a layer of solid agar of about 0·2 cm depth) is then poured into the dish alongside the suspension to minimize heat shock and then the suspension is thoroughly mixed with the still molten agar.

For this part of the operation replicate samples are invariably taken. Only experience with a particular water can guide the experimenter to minimize labour by choosing a restricted range of dilutions to submit to the plating procedure but yet give maximum accuracy at the final stage of colony counting. Step (3) incubate the plates in a selected atmosphere, at a selected temperature for a selected length of time. Then step (4) count the colonies which have appeared in or on (or both) the agar. Based on the count the number of living cells is calculated per unit volume of water. Viable cells of microorganisms can also be counted by use of the membrane filter. In this method colonies which have grown on the membrane filter which has been placed on the surface of nutrient agar medium are counted. The funnel used for the filtration of sample is sterilized with alcohol and flamed. The membrane filters (pore size 0·3—0·5μm) are sterilized by boiling in water. The amount of filtered water to be used depends on the expected number of micro-organisms. Usually 0·1—5 ml of the sample is used directly or after being diluted to 10 or 100 ml. By the filtration of a small volume of the sample 5 or 10 ml of sterile water have to be added into the funnel. After filtration the filters are picked up from the funnel and transferred to the surface of nutrient agar medium. The colonies are grown at room temperature for 5 days. Then the filters are dried and fixed in fumes of formaldehyde. In the laboratory the colonies on the filter surface are counted under the binocular lens. The colony count is facilitated by the previous staining with the erythrosin for 1—2 min, and subsequent washing with water (cf. 4.2). For counting micro-organisms in bottom sediments suspensions of the sediment in sterile water are used. The exact volume of the sediment sample is taken with a calibrated tube. Then 1—3 g of sample is dispersed by vigorous shaking in 250 ml of sterile water. The suspension thus prepared is subsequently diluted to the appropriate concentrations. Using the diluted samples a viable count on a membrane filter is made as described before. From the number of the colonies obtained the number of viable cells is recalculated per 1 g of the sediment.

4.8.2 The method of making dilutions.

Normal practice is to extract a measured volume of the suspension with a pipette and thoroughly mix this aliquot with a known volume of diluent. A minor source of error here can be caused by microbial cells adhering to the sides of the pipette or the diluent vessel and thus not being transferred eventually to the agar medium.

However a far greater potential source of error is attributable to the adsorption of micro-organisms on particulate matter. In waters which contain little particulate matter most of the micro-organisms may be free floating, but if waters which habitually contain particulate matter are to be examined routinely it is advisable to standardize the sampling procedure. The water should first be thoroughly shaken and then allowed to stand for a specified length of time, and then the sample should be taken carefully always from the same depth below the surface. (A clothes peg attached to the pipette provides a convenient marker.)

Aggregates of cells will behave like colonized particles and yield only one visible colony for the final count. It is not unusual to observe that plating of serial dilutions does not yield colony counts in proportion to the dilution. Increasing the dilution usually produces a disproportionately higher count. 'Crowding' of the colonies on the plates is one explanation but increasing dispersion of aggregates with increasing dilution can be another explanation.

A careful study of the survival of non-growing cells of *Aerobacter aerogenes* has shown that even minor modifications in the composition of the suspending fluid can affect viability (Postgate and Hunter, 1962). Several commonly used diluents have been shown to have a marked bactericidal effect on certain bacteria (King and Hurst, 1963). Use of the same water, sterilized, is probably the best compromise. In waters, microbial growth may be limited by ions such as ammonium and phosphate ions and if these are included in diluent fluids in relatively high concentration they could lead to 'substrate-accelerated' death (Postgate and Hunter 1964). It is a common microbiological practice to store fluids and materials at refrigerator temperatures but the precaution of equilibrating sample and diluent liquids to the same temperature is not always taken. Cold shock is an established phenomenon (Strange and Dark, 1962) and recent work has shown that psychrophilic organisms are not entirely immune to this deleterious effect (Farrell and Rose, 1968). At the stage where the diluted sample is brought into contact with the growth medium, trauma from abrupt temperature change can be avoided in the 'surface drop' or 'spread plate' techniques but not in the poured plate one.

One of the basic assumptions of the plate count is that each microbial cell produces a visible colony. This can be realized only when the average distance between colonies is large enough to prevent growth inhibition by metabolic products. On the other hand, the higher the number of colonies per plate, the lower the statistical error. Thus, between these limits there is an optimum density of colonies that will give the most reliable count.

Figure 26 (Melchiorri-Santolini, 1969) demonstrates that the inhibition effects become very strong when we have more than 50—60 colonies per plate.

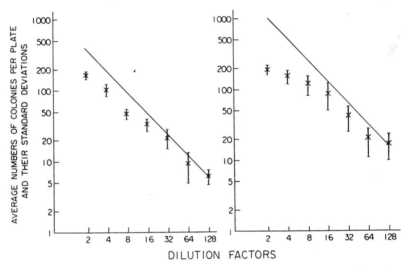

Figure 26. Log-log plots of the results of two experiments of plate counts of samples from Lake Maggiore. Dilution ratio 1:2, ten replicates per dilution; surface spread agar plates; peptone yeast extract medium; six days incubation at 24°C. The straight line represents the attended number of colonies on the basis of the results obtained from the more diluted plates.

Exceeding this density leads to a loss of reasonable accuracy. At low colony densities, the stochastic error increases, even if the growth response is better. In order to keep the statistical error in reasonable limits the number of colonies per plate should not be less than 15—20.

In order to obtain plates with an average number of colonies within these limits, the usual dilution ratio 1:10 is not sometimes suitable. A dilution ratio 1:4 or even less has to be used.

4.8.3 Nature of the plating medium and the condition of incubation.

For natural sources the viable count is usually only a fraction of the total count obtained by direct microscopical observations. Although the discrepancy may be partly attributable to the fact that many of the organisms observed directly are non viable, nevertheless it is likely that no conceivable medium and condition of incubation would be conductive to the growth of all viable organisms in the sample. This likelihood is the basis of the technique of selective and elective culture. For selective culture the aim is to inhibit the growth of organisms other than those of the group required and for elective culture the aim is to construct a medium which will allow the group required to dominate the population which develops. Although studies of the nutrition of aquatic micro-organisms are sparse, there is much evidence that gram-negative bacteria, classed as *Pseudomonas*, *Achromobacter*, *Moraxella* or *Acinetobacter* species constitute an appreciable part of the bacterial flora of waters (Collins, 1963; Baumann, 1968) and generally these organisms have wide catabolic powers and rather simple growth requirements. The composition of the media usually used includes beef or fish broth, peptone, glucose, yeast extract, ferrous phosphate, salts of nitrate, ammonium and phosphate (Hayes and Anthony, 1959; Jannasch, 1967; Jannasch and Jones, 1959). Simple commercial beef-peptone media are also usually used for the plate count of most heterotrophic micro-organisms. These media are made for the sanitary analysis of water and sewage, and too concentrated for the plate count of the purely aquatic micro-organisms. The appropriate media for this purpose is the solid agar medium containing 0·5—1% of dehydrated preparation and 1·5% of agar. A reduction in the strength of the nutrient components of common nutrient agars to one-fifth or even one-tenth normal strength yields a higher count than the normal agar. This phenomenon may be related to the fact that the osmotic pressure of many waters is low. Hence, of those organisms which enter waters, only those which can rapidly adapt to a low osmotic pressure will retain their viability. If there exists a true indigenous microbial flora of waters which can only exist and multiply under conditions of very low nutrient level and is inhibited by high concentrations of nutrients then it cannot be detected by conventional means. A visible colony can only be produced if there is enough food available to allow the aggregate of cells to become sufficiently large. If an agar medium contains concentrations of nutrient adequate for one fraction of the population but inhibitory to another fraction the latter will not appear. However, if a concentration of nutrient conducive to the growth of the latter fraction is employed this may be too low to yield visible colonies of any species.

The optimum growth temperature of many of the micro-organisms found in natural waters, in temperate regions at least is much below 37°C and indeed such waters harbour a high proportion of psychrophilic bacteria (Baig and Hopton, 1969; Stokes and Redmond, 1966). The nature of the medium can influence the optimum and maximum temperature of growth but nevertheless an incubation temperature of 20°—25° C is almost always likely to be preferable to 37° C. Incubation periods of 2—5 days are usually reasonable. The colonies can be counted twice after 2 and 5 days of incubation.

4.9 Estimation of microbial concentration in dilution series (MPN)

The method of dilution series for the estimation of micro-organism densities is based on the assumption that in a tube, containing a suitable cultural medium, organisms may develop even if the tube is inoculated with a single viable cell. However, as it is in any case impossible to identify tubes inoculated with one or more cells, all we can do is merely to check the tubes in which growth has occurred. This means that the calculations must be exclusively based on the frequencies of the sterile tubes which represent therefore the event zero.

The probability (P) that a tube receives x cells when inoculated with one ml of a suspension containing m organisms/ml is

$$P = m^x \frac{1/\exp(m)}{x^1}, \text{ so that for } x = 0 \; P = 1/\exp(m)$$

This method is commonly used in aquatic microbiology and usually gives higher numbers in comparison with plate counts (Jannasch and Jones, 1959; Melchiorri-Santolini, 1966). The main disadvantage of this method lies in the fact that we are dealing with a probabilistic estimation, with a high error.

This error can be reduced either by increasing the number of replicates (*i.e.* tubes) in each dilution or reducing the intervals of dilutions. The second method is by far the most convenient.

The precision of the method reaches its maximum when the percentage of the event zero (sterile tubes) is 20%. If, however, the percentage of sterile tubes lies in a range between 8 and 45%, the error is not very high (Fig. 27).

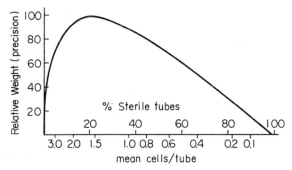

Figure 27. Relative precision of the bacterial count in the dilution series (modified from Peto 1953).

Table 3. Mean standard error of the estimated number of bacteria as a function of the dilution ratio (after Cavalli-Sforza, 1961).

Dilution ratio	Mean standard error in % of the estimated number
2	$65/\sqrt{n}$
3·16	$83/\sqrt{n}$
4	$91/\sqrt{n}$
8	$112/\sqrt{n}$
10	$120/\sqrt{n}$

Table 4. Mean standard error of the estimated number of bacteria in dilution series as dependent upon the number of parallel tubes and upon the dilution ratio (after U. Melchiorri-Santolini).

Dilution ratio	Number of tubes in each dilution	Number of tubes to cover a 1 to 100 range	Mean standard error—%
1:10	3	9	69
1:10	5	15	53
1:3·16	3	15	48
1:3·16	5	25	37
1:3·16	10	50	26
1:3·16	20	100	18

The dilution series should therefore include at least one dilution with such a concentration allowing it to be included in this range. As a consequence, the dilution ratio 1:10 is undoubtedly too wide. This is clearly demonstrated by Table 3 (Cavalli-Sforza, 1961).

For the necessary precision, the following methods are more convenient (Table 4).

The dilution 1:2 becomes convenient when the number of tubes in each dilution is very high.

The calculation method suggested by Peto (1953) seems to be the most appropriate and it is summarized as follows. (1) A provisional value of the number m_0 of bacteria/ml in the first dilution is drawn from any table giving values for 1:10 dilutions, and/or smaller number of tubes. (2) The number of bacteria/ml (m_i) proportional to m_0 in the dilution series is then reckoned. The m_i values for each dilution are added and the results multiplied for the number of tubes corresponding to each dilution: $S\, n_i\, m_i$.

(3) The values $A = \dfrac{m_i}{(1 - 1/\exp(m_i))} \quad B = \dfrac{m_i^2 \exp(m_i)}{(\exp(m_i) - 1)^2}$

are then calculated or found in Tables 5 and 6.

(4) Two sums are made: $SA_i f_i$ and $SB_i f_i$, where f_i is the number of *fertile* tubes for each dilution.

(5) The new estimation m_1 is obtained through the following formula:

$$m_1 = m_0 \cdot \left(1 + \frac{S A_i f_i - S n_i m_i}{S B_i f_i}\right)$$

(6) If m_1 is very different from m_0, a new cycle of calculations must be run, using m_1 as the provisional value.

(7) The variance of this estimation is $\dfrac{m_1^2}{S B_i f_i}$.

(8) The confidence limits at 95% are then obtained:

$$m_i \pm 1.96 \sqrt{\left(\frac{m_1^2}{S B_i f_i}\right)}, \quad 1.96 \text{ being the value for}$$
$p = 0.05$

of the probability integral. For different levels of probability, take the value of 'Student's' t for ∞ degrees of freedom.

Table 5. The value $A = \dfrac{m_i}{(1 - 1/\exp(m_i))}$ (After S. Peto)

m_i	0.00	0.01	0.02	0.03	0.04	0.05	0.06	0.07	0.08	0.09
.0	1.000	1.005	1.010	1.015	1.020	1.025	1.030	1.035	1.041	1.046
.1	1.051	1.056	1.061	1.066	1.072	1.077	1.082	1.087	1.093	1.098
.2	1.103	1.109	1.114	1.119	1.125	1.130	1.136	1.141	1.147	1.152
.3	1.157	1.163	1.169	1.174	1.180	1.185	1.191	1.196	1.202	1.208
.4	1.213	1.219	1.225	1.230	1.236	1.242	1.248	1.253	1.259	1.265
.5	1.271	1.277	1.282	1.288	1.294	1.300	1.306	1.312	1.318	1.324
.6	1.330	1.336	1.342	1.348	1.354	1.360	1.366	1.372	1.378	1.384
.7	1.391	1.397	1.403	1.409	1.415	1.421	1.428	1.434	1.440	1.446
.8	1.453	1.459	1.465	1.472	1.478	1.485	1.491	1.497	1.504	1.510
.9	1.517	1.523	1.530	1.536	1.543	1.549	1.556	1.562	1.569	1.575
1.0	1.582	1.589	1.595	1.602	1.609	1.615	1.622	1.629	1.635	1.642
1.1	1.649	1.656	1.662	1.669	1.676	1.683	1.690	1.697	1.703	1.710
1.2	1.717	1.724	1.731	1.738	1.745	1.752	1.759	1.766	1.773	1.780
1.3	1.787	1.794	1.801	1.808	1.815	1.822	1.830	1.837	1.844	1.851
1.4	1.858	1.865	1.873	1.880	1.887	1.894	1.902	1.909	1.916	1.924
1.5	1.931	1.938	1.946	1.953	1.960	1.968	1.975	1.982	1.990	1.997
1.6	2.005	2.012	2.020	2.027	2.035	2.042	2.050	2.057	2.065	2.072
1.7	2.080	2.088	2.095	2.103	2.110	2.118	2.126	2.133	2.141	2.149
1.8	2.156	2.164	2.172	2.180	2.187	2.195	2.203	2.211	2.219	2.226
1.9	2.234	2.242	2.250	2.258	2.266	2.273	2.281	2.289	2.297	2.305
2.0	2.313	2.321	2.329	2.337	2.345	2.353	2.361	2.369	2.377	2.385
2.1	2.393	2.401	2.409	2.417	2.425	2.433	2.442	2.450	2.458	2.466
2.2	2.474	2.482	2.490	2.499	2.507	2.515	2.523	2.532	2.540	2.548
2.3	2.556	2.565	2.573	2.581	2.589	2.598	2.606	2.614	2.623	2.631
2.4	2.639	2.648	2.656	2.665	2.673	2.681	2.690	2.698	2.707	2.715
2.5	2.724	2.732	2.741	2.749	2.757	2.766	2.774	2.783	2.792	2.800
2.6	2.809	2.817	2.826	2.834	2.843	2.851	2.860	2.869	2.877	2.886
2.7	2.895	2.903	2.912	2.920	2.929	2.938	2.946	2.955	2.964	2.973
2.8	2.981	2.990	2.999	3.007	3.016	3.025	3.034	3.043	3.051	3.060
2.9	3.069	3.078	3.086	3.095	3.104	3.113	3.122	3.131	3.139	3.148
3.0	3.157	3.166	3.175	3.184	3.193	3.202	3.211	3.219	3.228	3.237
3.1	3.246	3.255	3.264	3.273	3.282	3.291	3.300	3.309	3.318	3.327
3.2	3.336	3.345	3.354	3.363	3.372	3.381	3.390	3.399	3.408	3.417
3.3	3.426	3.435	3.445	3.454	3.463	3.472	3.481	3.490	3.499	3.508
3.4	3.517	3.527	3.536	3.545	3.554	3.563	3.572	3.581	3.591	3.600
3.5	3.609	3.618	3.627	3.637	3.646	3.655	3.664	3.673	3.683	3.692
3.6	3.701	3.710	3.720	3.729	3.738	3.747	3.757	3.766	3.775	3.785
3.7	3.794	3.803	3.812	3.822	3.831	3.840	3.850	3.859	3.868	3.878
3.8	3.887	3.896	3.906	3.915	3.924	3.934	3.943	3.952	3.962	3.971
3.9	3.981	3.990	3.999	4.009	4.018	4.028	4.037	4.046	4.056	4.065
4.0	4.075	4.084	4.093	4.103	4.112	4.122	4.131	4.141	4.150	4.160
4.1	4.169	4.179	4.188	4.198	4.207	4.216	4.226	4.235	4.245	4.254
4.2	4.264	4.273	4.283	4.292	4.302	4.312	4.321	4.331	4.340	4.350
4.3	4.359	4.369	4.378	4.388	4.397	4.407	4.416	4.426	4.436	4.445
4.4	4.455	4.464	4.474	4.483	4.493	4.503	4.512	4.522	4.531	4.541
4.5	4.551	4.560	4.570	4.579	4.589	4.599	4.608	4.618	4.627	4.637
4.6	4.647	4.656	4.666	4.676	4.685	4.695	4.705	4.714	4.724	4.733
4.7	4.743	4.753	4.762	4.772	4.782	4.791	4.801	4.811	4.820	4.830
4.8	4.840	4.850	4.859	4.869	4.879	4.888	4.898	4.908	4.917	4.927
4.9	4.937	4.946	4.956	4.966	4.976	4.985	4.995	5.005	5.014	5.024

Table 6. The value $B = \dfrac{m_i^2 \exp(m_i)}{(\exp(m_i) - 1)^2}$ (After S. Peto)

m_i	0.00	0.01	0.02	0.03	0.04	0.05	0.06	0.07	0.08	0.09
.0	1.000	1.000	1.000	1.000	1.000	1.000	1.000	1.000	0.999	0.999
.1	0.999	0.999	0.999	0.999	0.998	0.998	0.998	0.998	0.997	0.997
.2	0.997	0.996	0.996	0.996	0.995	0.995	0.994	0.994	0.993	0.993
.3	0.993	0.992	0.992	0.991	0.990	0.990	0.989	0.989	0.988	0.987
.4	0.987	0.986	0.985	0.985	0.984	0.983	0.983	0.982	0.981	0.980
.5	0.979	0.979	0.978	0.977	0.976	0.975	0.974	0.973	0.972	0.971
.6	0.971	0.970	0.969	0.968	0.967	0.966	0.964	0.963	0.962	0.961
.7	0.960	0.959	0.958	0.957	0.956	0.954	0.953	0.952	0.951	0.950
.8	0.948	0.947	0.946	0.945	0.943	0.942	0.941	0.939	0.938	0.937
.9	0.935	0.934	0.932	0.931	0.930	0.928	0.927	0.925	0.924	0.922
1.0	0.921	0.919	0.918	0.916	0.915	0.913	0.911	0.910	0.908	0.907
1.1	0.905	0.903	0.902	0.900	0.898	0.897	0.895	0.893	0.892	0.890
1.2	0.888	0.886	0.885	0.883	0.881	0.879	0.878	0.876	0.874	0.872
1.3	0.870	0.868	0.867	0.865	0.863	0.861	0.859	0.857	0.855	0.853
1.4	0.852	0.850	0.848	0.846	0.844	0.842	0.840	0.838	0.836	0.834
1.5	0.832	0.830	0.828	0.826	0.824	0.822	0.820	0.818	0.816	0.814
1.6	0.811	0.809	0.807	0.805	0.803	0.801	0.799	0.797	0.795	0.792
1.7	0.790	0.788	0.786	0.784	0.782	0.780	0.777	0.775	0.773	0.771
1.8	0.769	0.767	0.764	0.762	0.760	0.758	0.755	0.753	0.751	0.749
1.9	0.747	0.744	0.742	0.740	0.738	0.735	0.733	0.731	0.729	0.726
2.0	0.724	0.722	0.720	0.717	0.715	0.713	0.710	0.708	0.706	0.704
2.1	0.701	0.699	0.697	0.694	0.692	0.690	0.687	0.685	0.683	0.681
2.2	0.678	0.676	0.674	0.671	0.669	0.667	0.664	0.662	0.660	0.657
2.3	0.655	0.653	0.651	0.648	0.646	0.644	0.641	0.639	0.637	0.634
2.4	0.632	0.630	0.627	0.625	0.623	0.620	0.618	0.616	0.614	0.611
2.5	0.609	0.607	0.604	0.602	0.600	0.597	0.595	0.593	0.590	0.588
2.6	0.586	0.584	0.581	0.579	0.577	0.574	0.572	0.570	0.568	0.565
2.7	0.563	0.561	0.559	0.556	0.554	0.552	0.549	0.547	0.545	0.543
2.8	0.540	0.538	0.536	0.534	0.532	0.529	0.527	0.525	0.523	0.520
2.9	0.518	0.516	0.514	0.512	0.509	0.507	0.505	0.503	0.501	0.498
3.0	0.496	0.494	0.492	0.490	0.488	0.485	0.483	0.481	0.479	0.477
3.1	0.475	0.473	0.470	0.468	0.466	0.464	0.462	0.460	0.458	0.456
3.2	0.454	0.452	0.449	0.447	0.445	0.443	0.441	0.439	0.437	0.435
3.3	0.433	0.431	0.429	0.427	0.425	0.423	0.421	0.419	0.417	0.415
3.4	0.413	0.411	0.409	0.407	0.405	0.403	0.401	0.399	0.397	0.395
3.5	0.393	0.391	0.389	0.388	0.386	0.384	0.382	0.380	0.378	0.376
3.6	0.374	0.372	0.371	0.369	0.367	0.365	0.363	0.361	0.359	0.358
3.7	0.356	0.354	0.352	0.350	0.349	0.347	0.345	0.343	0.342	0.340
3.8	0.338	0.336	0.334	0.333	0.331	0.329	0.328	0.326	0.324	0.322
3.9	0.321	0.319	0.317	0.316	0.314	0.312	0.311	0.309	0.307	0.306
4.0	0.304	0.302	0.301	0.299	0.298	0.296	0.294	0.293	0.291	0.290
4.1	0.288	0.286	0.285	0.283	0.282	0.280	0.279	0.277	0.276	0.274
4.2	0.273	0.271	0.270	0.268	0.267	0.265	0.264	0.262	0.261	0.259
4.3	0.258	0.256	0.255	0.254	0.252	0.251	0.249	0.248	0.246	0.245
4.4	0.244	0.242	0.241	0.239	0.238	0.237	0.235	0.234	0.233	0.231
4.5	0.230	0.229	0.227	0.226	0.225	0.223	0.222	0.221	0.220	0.218
4.6	0.217	0.216	0.215	0.213	0.212	0.211	0.210	0.208	0.207	0.206
4.7	0.205	0.203	0.202	0.201	0.200	0.199	0.197	0.196	0.195	0.194
4.8	0.193	0.192	0.190	0.189	0.188	0.187	0.186	0.185	0.184	0.183
4.9	0.181	0.180	0.179	0.178	0.177	0.176	0.175	0.174	0.173	0.172

(9) The estimation of the probability of the results observed can be made with the χ^2 test, calculating the expected frequency of the number of sterile tubes in each dilution with the formula $1/\exp(m)$, where m is the number of bacteria/ml calculated in each dilution. The χ^2 shall have a number of degrees of freedom corresponding to the number of dilutions minus 1.

(10) The χ^2 has to be calculated, in this case, by applying the formula:

$$\frac{(F_t - F_O)^2}{F_t (1 - S)}$$

(Peto, 1953), where $S = 1/\exp(m)$, F_t is the theoretical frequency of sterile tubes, that is $S \cdot n$ (n = number of tubes in each dilution), F_0 is the observed frequency of sterile tubes, $S = 1/\exp(m)$ is the probability that a tube remains sterile when inoculated with 1 ml of the suspension containing m organisms/ml. It is worthwhile to point out that the application of this test is sometimes not very correct, since the value of the expected frequencies is often too small, but, when the results of many experiments are considered together, it is useful for testing whether they correspond to the frequencies which should be expected. Should this method give results with low probability, there are two possible interpretations. A contamination of the more diluted tubes or a breakage into single cells of bacterial aggregates during the dilutions. This event will evidently result in an underestimation of the real number of bacterial cells present in the sample.

It is to be stressed that this method gives in any case a very good estimation of the bacterial densities, but the estimation of variance and consequently of confidence limits becomes too imprecise when the number of tubes for each dilution is less than ten. In this case, it is convenient to calculate the standard error of $\log_{10}$ of the Most Probable Number (Cochran, 1950). This transformation is necessary because of the strong skewness of the distribution of the estimated densities in experiments with few tubes for each dilution (*i.e.* less than ten).

The formula for the calculation of the standard error of $\log_{10}$ of the MPN is

$$0.55 \cdot \sqrt{\frac{\log_{10} a}{n}}$$

where a is the dilution ratio and n is the number of tubes for each dilution (with a dilution ratio of 10 use a factor of 0·58 instead of 0·55).

To test the significance of the difference between two estimations obtained from independent series with small numbers of tubes for each dilution, a 'Student' t (degrees of freedom $= \infty$) can be computed:

$$t = \frac{\log d_1 - \log d_2}{0 \cdot 55 \cdot \sqrt{\dfrac{\log a_1}{n_1} + \dfrac{\log a_2}{n_2}}}$$

where d_1 and d_2 are the two estimations, a_1 and a_2 are the dilution ratios and n_1 and n_2 are the numbers of tubes for each dilution. With a dilution rate of 10 a factor of 0·58 instead of 0·55 is to be used and referred to the normal probability tables.

And finally a practical suggestion. All the calculation methods for the estimation of bacterial concentration in dilution series are based on the assumption that a certain volume of water is taken directly from the original sample. This is never done and successive dilutions of the sample are used instead. If the dilution volumes are too small, a stochastic error is introduced in the experiment, unless the dilution volumes are at least three times larger than the minimum required.

4.10 Relation between cultural and microscopic counts of micro-organisms

The ratio between numbers of micro-organisms counted as colonies on agar plates and direct microscopic counts vary with the type of water investigated and with the type of medium used. Assuming that the microscopic count of micro-organisms is proportional to the amount of readily decomposable organic substances in water (Straskrabova, 1968), it can serve as a rough estimate of the degree of pollution. Comparisons between plate counts and direct counts may result in new information only when waters of similar type and with similar direct counts are studied (Table 7).

4.11 Application of different methods of enumeration of micro-organisms

All methods of microbial counts have their limitations and advantages. The methods to be used should be selected bearing in mind the purpose of the

Table 7. The percentage of plate count in direct membrane filter count in different waters.*
(After V. Straskrabova)

Locality	BOD$_5$	Direct counts (1000/ml)	$\dfrac{PA}{\text{direct count}} \cdot 100$ %	$\dfrac{BPA}{\text{direct count}} \cdot 100$ %
Klicava reservoir	2·7	425	0·87	0·34
Vltava river	4·0	1900	4·30	1·16
Kocaba brook	2·3	678	12·4	3·22
Ledinice brook	5·5	1160	13·5	13·2
Raw sewage	380	126000	—	12·2
Treated sewage	32	16300	—	22·2

* BOD$_5$—biochemical oxygen demand in 5 days, mg O$_2$/l. PA—colonies on poor agar (0·5g bactopeptone per 1 litre; incubated at 20°C for 20 days). BPA—colonies on an ordinary beef-pepton agar, incubated at 20°C for 10 days. The values of surface waters are the means of at least 10 samples taken during the whole year, the analyses of sewage were made from September to December.

study. The plate count or MPN methods can be used in cases where information on the relative abundance of heterotrophic micro-organisms in different aquatic environments is desired. These methods can also be employed when the number of micro-organisms of different physiological types is needed.

The direct microscopic count is appropriate when the biomass of planktonic, benthic and periphytonic micro-organisms has to be estimated in production studies. A close inter-relation is usually found between the level of productivity of a water body and the total count or biomass of the micro-organisms as measured by the direct microscopic method (Table 8). This method is also useful in the evaluation of micro-organisms as the food for secondary producers.

4.12 Total microbial biomass estimation by the ATP method

4.12.1 Principle. The determination of biomass (the total amount of living cells) in either fresh water or in seawater is complicated by the presence of non-living, detrital material. In order to estimate the biomass in any sample, it is necessary either to make a complete separation of these two fractions or to analyse for some constituent that is found in live cells but not in detrital material. As there is a complete overlap in size distribution of living cells and

Table 8. Number and biomass of bacteria in water bodies of different trophic levels (After Y. Sorokin).

Type of water body	Water			Bottom sediments	
	Total number, 10^6/ml	Wet weight of biomass, (g/m^3)	Ratio between total count and plate count	Total number, 10^9/g wet sediment	Wet weight of biomass g/kg wet sediment
Waste waters and polluted basines	10–100	15–150	10–50	10–20	10–20
Eutrophic	2–10	1·5–15	50–500	2–10	1·5–10
Mesotrophic	0·5–2	0·2–1·5	500–2000	0·2–2	0·15–1·5
Oligotrophic	0·1–0·5	0·02–0·2	2000–10000	0·05–0·2	0·01–0·15

Determination of Microbial Numbers and Biomass

detrital particles, and also because many cells are adsorbed or attached to detritus, there is no feasible way to affect a separation of these two fractions. If a cellular constituent is to serve as a biomass indicator, it must be found in fairly uniform concentrations in all living cells, and must not be associated with non-living, particulate material. Adenosine triphosphate (ATP) fulfils these criteria and is also a very practical constituent to use as very small amounts can be measured by use of the bioluminescent reaction occurring in firefly tails. The essential part of this reaction in regard to the bioassay method is that one photon of light is emitted for each molecule of ATP which is hydrolysed. The concentration of ATP in any sample can thus be obtained by measuring the intensity of the light emission when the sample is mixed with the proper reactants and enzyme. As the ATP level in microbial cells is fairly uniform, the ATP concentration in any sample can be used to calculate total biomass in terms of organic carbon or dry weight.

4.12.2 Outline of procedure.

a. Obtaining of sample.
In aquatic environments where the amount of living material is very low between 1 and 2 litres of sample are required. For such environments, either sterile samplers must be employed, or devices such as Van Dorn bottles must be carefully cleaned before use. In more productive waters, in which sample volumes of less than 100 ml may often be employed, the type of sampling device is not so critical. The ATP method of biomass estimation has not as yet been employed for studies on sediments or consolidated materials (such as activated sludge in sewage treatment), but a few grams of such material should normally yield sufficient ATP for a biomass estimation.

b. Treatment of sample.
Water samples may be poured directly into clean pyrex glass bottles, or the samples may be poured through a coarse filter to remove all materials larger than any desired size. The samples should be used as soon as possible after collection to minimize changes occurring in the cells, but storage periods of up to one hour in the dark seem to be all right. The sample is then poured through a filter which will retain all the particulate material in which one is interested. For microbial studies in natural waters, 47 mm diameter membrane filters with a pore size of $0.45 \mu m$ are generally used, although other filters (such as glass fibre filters) may also be employed depending on the size

distribution of the cells in the sample. This filtration is done at reduced pressures of about 1/3 atmosphere, as higher vacua may have deleterious effects on many cells. As soon as the filtration is complete, the filter is removed from the glass filter unit and immersed in 5·0 ml of boiling Tris buffer (0·02 M, pH 7·75). It is very important that the temperature of the Tris buffer be close to 100° C and that the elapsed time between the completion of filtration to the time at which the entire filter is immersed in the Tris buffer be as short as possible. The test tube containing the buffer and filter is then kept in the boiling water bath for five minutes to kill all cells and to extract all cellular ATP, after which the tube is removed, cooled, capped, and then frozen until the time of analysis. If sediments or sludge is to be tested for ATP, the filtration step is omitted, and measured or weighed amounts of materials are added directly to boiling Tris buffer. In these cases, however, it is important to have the volume of Tris buffer sufficiently large that the temperature does not go below 90° C upon addition of the sample, and that the salt content of the extract does not interfere with the enzymatic reaction (if it does, the extract must be diluted with more Tris buffer).

c. Enzyme preparation.
In order to have one photon of light emitted for each molecule of ATP which is hydrolysed, the other reactants (reduced luciferin, luciferase, Mg^{2+}, and oxygen) must be present in excess. Lyophilized water extracts from firefly lanterns (*Photinus pyralis*) to which have been added $MgSO_4$ and potassium arsenate are available commercially. These dry extracts are kept at about $-20°C$ until 10 hours before use, when they are hydrated with Tris buffer (0·02 M, pH 7·75) and kept at room temperature. After about 7 hours the turbid suspension is centrifuged for one minute and the supernatant is poured into a clean vial or tube. This solution is then left for another 2—3 hours at room temperature before use. The use of purified luciferase and luciferin would be preferable to the use of the crude lyophilized water extract.

d. Determination of ATP.
The frozen samples are thawed, taken to room temperature in a water bath, and then held on a tube buzzer immediately before use. The presence of particulate matter in the extract (either from cellular material or from dissolution of glass fibre filters) does not interfere with the ATP assay. Aliquots of the sample (usually 0·2 ml) are then added to an equal volume of the enzyme preparation and the amount of light emitted is measured. The

Determination of Microbial Numbers and Biomass 75

sample may be injected into the enzyme preparation while the latter is in close proximity to the light-detecting device, or the sample and enzyme preparation may be shaken by hand and then placed in front of the phototube after a set period of time (usually 15 seconds). The method which is used for the measurement of the emitted light depends upon the level of ATP in the sample. For quantities of ATP between a few µg to one mg or more, it is possible to use commercially available spectrophotometers or fluorometers with a few minor alterations. Either the peak height of light emission may be measured, or the integrated light intensity may be determined over a one- or two-minute time period. For sub-microgram quantities of ATP it is necessary to use more sensitive light detectors, and various scintillation counters may be found to be satisfactory (Addanki, Sotos, and Rearick, 1966; Cole, Wimpenny, and Hughes, 1967). For work in the range of 10^{-3} to 10^{-7} µg ATP, however, it is necessary to use a very sensitive photomultiplier tube. There are many ways to measure the emitted light with a photomultiplier tube, but the ways which have been found to be convenient are (1) to amplify the anode signal from the tube and then record it on a paper strip recorder, and (2) put the anode signal into a voltage-frequency converter and thence into an electronic counter. Details of the apparatus required for these determinations and the methodology involved may be found in the references by Holm-Hansen and Booth (1966), Strehler (1968), Strickland and Parsons (1968), and Lyman and De Vincenzo (1967).

4.12.3 Extrapolation of ATP values to biomass. A great variety of marine and freshwater bacteria and algae have been investigated to determine the cellular contents of ATP under a variety of environmental conditions (Hamilton and Holm-Hansen, 1967; Coombs, Halicki, Holm-Hansen, and Volcani, 1967; Holm-Hansen, 1969). The range of ATP concentrations found in these studies are from about 0·05 to 0·7% of the total cellular organic carbon value. Most of the ATP concentrations are clustered fairly close to the average value of 0·4% of the organic carbon value. To equate ATP values to cellular organic carbon, one therefore multiplies by the factor of 250. Assuming that the carbon content of most cells is 50% of the dry weight, it is necessary to multiply the ATP concentration by 500 to give one a biomass expressed in cellular dry weight. Data on ATP contents in bacteria may also be found in the papers by Cole, Wimpenny, and Hughes (1967) and D'Eustachio and Levin (1967).

4.12.4 Sensitivity and precision of method. The sensitivity of the above method is between 10^{-4} to 10^{-5} µg ATP (Holm-Hansen, 1969). This limit could be reduced considerably by using larger volumes of reactants. A higher sensitivity could also be obtained by using 'Biometer' of the DuPont Co., a compact and automatic ATP meter recently developed.

Ten replicate determination on algal cultures give standard deviations of about 5% from the mean (usually between 1 to 2×10^{-3} µg ATP). The largest errors in the entire procedure will come, however, in sampling errors of natural waters and in the factor equating ATP concentrations to biomass in terms of organic carbon or dry weight.

4.13 Evaluation of the aggregation level of planktonic bacteria

The value of bacteria as food for zooplankton in natural environments is strongly dependent upon the level of aggregation of their cells. Filtrators such as *Calanoida* can filter flakes of bacterial cells but not single cells (Sorokin, 1968, 1971). Therefore, the percentage of the aggregated cells in the total amount of bacteria (F) becomes one of the important characteristics of planktonic bacteria, when considering their trophic roles in the waters.

To measure the F value, a sample of water filtered through the finest plankton net is poured into a bottle. Then a solution of strongly labelled organic matter (cf. 5.7), or of labelled ^{14}C-carbonate (cf. 5.3) is added to the sample in darkness. After exposure for 1 day (or in the case of the use of ^{14}C-carbonate for 2 days) the sample is fixed with the formaldehyde. Then two duplicate 20—40 ml portions of the sample are filtered through membrane filters of 4—6µm pore size (Russian filter No. 6), and two other portions through the filters of 0·3µm pore size. In case of the labelled hydrolysate the filters are washed in the funnel by a subsequent filtration of 20 ml water (by portions), in case of the ^{14}C-carbonate they are washed with 1% HCl. The ratio of mean radioactivity of the coarse filters to that of the fine filters is expressed as a percentage value. It is used as a measure of the degree of aggregation of bacterial cells (F) in the water.

5
Estimation of Production Rate and *in situ* Activity of Autotrophic and Heterotrophic Micro-organisms

For the estimation of *in situ* production rate of micro-organisms in waters the following methods can be used: (a) estimation of rate of multiplication of micro-organisms as the rate of increase in number of microcolonies, growing on slides immersed in the water body; (b) direct measurement of increase in number of micro-organisms in isolated samples or on membrane filters; (c) estimation of the rate of multiplication of micro-organisms in continuous culture system; (d) calculation of production rate of micro-organisms based on the measurement of metabolic activities (the rate of autotrophic and heterotrophic CO_2 assimilation or the rate of O_2 consumption in isolated samples).

Among these methods b, c and d, which are widely used, are mentioned here.

5.1 Microcolony method

The planktonic micro-organisms are concentrated on membrane filters and incubated in contact with the natural water without addition of nutrients. The incubation is performed at the water temperature for several hours. Under these conditions it is known that the micro-organisms transferred into the changed conditions retain their own original reproduction rate for a limited period.

The water sample immediately after sampling is filtered through six parallel membrane filters. Three of them are placed on filter paper pads moistened with 3% aqueous solution of formaldehyde for determining the number of micro-organisms before incubation. The other three filters are placed in a Petri dish with a filter paper moistened with the natural water. Whatman paper for chromatography is appropriate for this purpose. The pads have to be boiled in distilled water, dried at 40—50° C and sterilized by autoclaving before use.

In case of unpolluted waters, an unfiltered water could be used for moistening the paper before incubation. When analysing polluted waters the water should be filtered through a membrane filter before use. During incubation the temperature of the sample and the filters should be kept at a temperature close to the *in situ* temperature.

After 3 hours' incubation the filters are placed on filter paper soaked with 3% formaldehyde, then dried and stained for counting.

The changes in microbial numbers on the filter ($\pm \Delta N$) is calculated as follows:

$$\pm \Delta N = \frac{N_f - N_i}{N_i}$$

where: N_f and N_i are microbial numbers after and before incubation respectively.

For estimation of the rate of reproduction of bacteria generation time (g) is calculated by the equation (Iwanoff, 1955):

$$g = \frac{t \cdot \lg 2}{\lg N_f - \lg N_i} \text{ hours}$$

where: t is incubation time in hours. When N decreases this value is negative and means the time for decrease the number to a half of the initial value.

The changes in number of micro-organisms on membrane filters coincides usually with those in bottles deprived of zooplankton.

5.2 Estimation of changes in number of bacteria in isolated water samples

The water sample is dispensed into two 250 ml bottles with or without preliminary treatment. One of the bottles is filled with the water which has previously been filtered through a membrane filter (pore size: 4—6μm) or fine mesh plankton net to exclude zooplanktons. The other bottle is filled with the water without previous filtration. At the beginning and at the end of the experiment the total number of bacteria in both the bottles are measured by the direct microscopic method (cf. 4.2). The bottles are exposed *in situ* in the water body or in similar conditions in the incubator (Rasumoff, 1947; Iwanoff, 1955).

The production of bacteria (expressed as number/ml) can be calculated by the following equations (Gak, 1969):

$$P = N_2 - N_1 + G \tag{1}$$

where: P = production of bacteria; N_1 = the initial number of bacteria in nonfiltered water; N_2 = the final number in the same water after the exposure for t hours; G = grazing of bacteria by zooplankton.

The value G can be obtained by the following equations (Romanova and Zonoff, 1964):

$$G = \frac{n_1 \cdot \exp(Kt) - N_2}{\exp(Kt) - 1} \cdot K \cdot t, \tag{2}$$

$$K = \frac{2{,}303 \; (\lg n_2 - \lg n_1)}{t} \tag{3}$$

where: n_1 and n_2 = the numbers of bacteria in the previously filtered water in the bottle before and after exposure for t hours; K = coefficient of the rate of bacterial multiplication. The equation (2) can be simplified (Gak, 1967) as follows:

$$G = \frac{+ Kt(n_1 - N_2)}{\dfrac{n_1}{N_1} - 1} \tag{4}$$

The biomass is calculated from the microbial number (cf. 4.3). The estimation of production rate of bacteria can be simplified if we assume that the increase in number by multiplication is equal to the decrease due to the grazing by zooplankton ($P = G$), e.g., the number of bacteria in the unfiltered natural water does not change during the experiment ($N_1 = N_2$). Such a case is often found in the reservoir water (Gak, 1971). According to the data obtained by Gak (1970) with 260 samples of the nonfiltered natural water, the number of bacteria did not actually change. The difference between N_1 ($3 \cdot 78 \pm 0 \cdot 19 \cdot 10^6$ cells/ml), and N_2 ($3 \cdot 68 \pm 0 \cdot 19 \cdot 10^6$ cells/ml) was not statistically reliable. These data prove that in the natural ecosystem the production of bacteria to a certain extent is compensated by the grazing by zooplanktons. If $P = G$ and $N_1 \pm N_2 = \bar{N}$, the production of bacteria during t time, P will be equal to:

$$P = \bar{N} \cdot K \cdot t \qquad (5)$$

where: N is average number of bacteria in the unfiltered sample. If the equation (5) is divided by N and the values N and P are expressed as biomass (cf. 4.3) the following formula can be obtained:

$$K \cdot t = \frac{P}{B} \qquad (6)$$

This means that $K \cdot t$ is equal to the coefficient P/B. In order to estimate the production of bacteria using the formula (5) we have to estimate the total number of bacteria in the unfiltered sample water (N) and the values of n_1 and n_2 with the previously filtered water of the same sample. Time of exposure of the bottles has to be selected so that the number of bacteria increases twice during the exposure. The values n_1 and n_2 can be replaced by the values H_1 and H_2 (cf. equation 7).

Using the simplified formula (5) recalculation can be done from previously published data on the bacterial production which had been calculated by use of the insufficient formula by Iwanoff (1955). The value of production (P) obtained by the formula (5) is equal to that obtained by Iwanoff's formula multiplied by 1·44 (Gak, 1971).

Stability of the number of bacteria N during the bottle experiment with the unfiltered water has theoretically a statistical regularity (Gak, 1971). But this is not the case with actual experiments. In some cases the number of bacteria increases ($G < P$), and in the other cases it decreases ($G > P$). The theoretical calculations and the actual experimental data are shown in Fig. 28. During 24 hours exposure the limits of ratio $\frac{N_1}{N_2}$ is between 0·5 and 4, e.g. 4 times by the $K \cdot t$-values between 0 and 1 (usual for the natural water). The error in the calculation of $P:B$ with the aid of the simplified formula (6) does not exceed 5% of that when the value of P is calculated according to the complete formula (2,3) by Romanova and Zonov (1964).

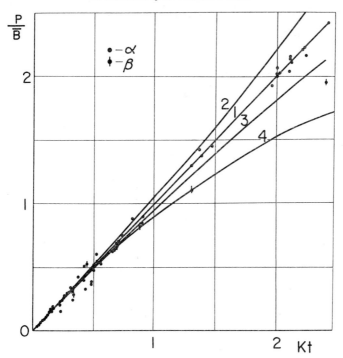

Figure 28. Relation between $\frac{P}{B}$ and K.t at the different initial (N_1) and final (N_2) numbers of bacteria in theory and in bottle experiments with the unfiltered water of Dnieper Reservoirs. Theoretical curves:

1—when $\frac{N_2}{N_1} = 1$; 2—when $\frac{N_2}{N_1} = 0.5$;

3—when $\frac{N_2}{N_1} = 2$; 4—when $\frac{N_2}{N_1} = e^{k \cdot t}$

Experimental data: α—when $0.5 \leqslant \frac{N_2}{N_1} \leqslant 2$; β—when $2 \leqslant \frac{N_2}{N_1} \leqslant e^{k \cdot t}$.

For the assessment of the time of one generation (g) Romanenko (1969) offers the method based upon the use of values of the production of bacteria (P) per hour as measured by use of the ^{14}C method, or of the values of the assimilation of CO_2 by microflora (H_1 and H_2) estimated twice during the time interval (t hours) (cf. 5.3). The calculations are made according to the following formula

$$g = \frac{t \lg 2}{\lg(B+P) - \lg B} = \frac{t \lg 2}{\lg H_2 - \lg H_1} \text{ hours} \qquad (7)$$

where: B = the biomass of planktonic bacteria, as measured by direct microscopy (cf. 4.3). Instead of the values H_1 and H_2 the corresponding values of the radioactivity of bacteria on the filters can be used.

Experimental data show that the generation time (g) depends upon the exposure time. The most common finding is the increase of g value with the increase of the time of exposure. It depends probably upon the uneven rate of multiplication of a different part of the bacterial population in the isolated water sample. Thus the exposure during this experiment has to be not less than 24 h, for at the shorter exposures the fluctuations of g values increase. It is important, therefore, to point out the time of exposure used.

The estimation of P by the use of the above formula is not simple, especially if there are a lot of stations to be studied in large water bodies. Moreover, the direct count method in some cases, especially in oligotrophic waters poor in zooplankton, does not permit the detection of grazing. In some cases it is possible to use a simplified method of evaluation of the bacterial production and its grazing by zooplankton. According to this method the difference between the initial and the final concentrations of bacteria in the bottles with the filtered water is measured. It gives a direct measure of increase of the number of bacteria or their production. The grazing can be more precisely measured in special grazing experiments during which a definite number of animals capable of filter feeding are placed into a definite volume of the same water and the decrease in the number of bacteria grazed by animals is measured. Based on the rate of grazing of bacteria by animals predominating in the water body and also their number in the water body it is possible to calculate the *in situ* grazing.

5.3 Estimation of production of heterotrophic bacteria using ^{14}C

5.3.1 Introduction. The autotrophic as well as the heterotrophic biosynthesis of bacteria proceed by involving external CO_2. By use of the ^{14}C method it is possible to measure not only photosynthesis by algae but also bacterial assimilation of CO_2. The microbial assimilation of CO_2 in water bodies is a result of the metabolic activity of different groups of micro-organisms which assimilate CO_2 in different amounts (Sorokin, 1965; Romanenko, 1965).

Heterotrophic micro-organisms which decompose proteins generally use 3—5% of the carbon of external CO_2 for the biosynthesis of cells. This group of bacteria contributes mainly to the dark assimilation of CO_2 which can be detected in the surface layer of the eutrophic and mesotrophic lakes and in the whole water mass of oligotrophic water bodies.

Bacteria having an intermediary type of metabolism between true heterotrophs and true chemoautotrophs, on the other hand, oxidize low molecular weight compounds produced during the anaerobic decay of organic matter (methane, methanol, formic acid) and assimilate 30—90% of the external CO_2 for the cell biosynthesis.

The chemoautotrophic bacteria which synthesize all the substances of their cells from the external CO_2, use the energy derived from the oxidation of inorganic reduced substances such as H_2, H_2S, NH_4, and Fe^{++}.

The latter two groups are the most active in the boundary layers between aerobic and anaerobic zones in the water column or in the interface layers between overlying water and bottom sediments.

The above mentioned differences in the relative amounts of CO_2 uptake by different types of micro-organisms makes the interpretation of the data on the dark uptake of CO_2 as a measure of the bacterial biosynthesis in a water body quite difficult. Nevertheless experience in studies of bacterial production and metabolism *in situ* conditions using the bottle technique shows that this determination can provide an approximation in solving of these problems. It is generally found that in waters having no direct contact with the anaerobic zones there is a close relationship between dark uptake of CO_2 measured by use of $^{14}CO_2$ (A) and the production of microbial biomass (P): A 0.06 P (Romanenko, 1964). According to Overbeck, this ratio is: $A = 0.066 P$, and according to Sorokin (1970): $A = 0.05 P$. Using this ratio it is possible to estimate roughly the production of heterotrophic bacteria in waters having no direct contact with the anaerobic zones if it is assumed that in these waters the dark uptake of CO_2 proceeds mostly through the metabolism of heterotrophic bacteria. The practical estimations showed that this assumption can be accepted as a first approximation (Romanenko, 1963, 1964 b). For the estimation of bacterial production, the formula: $P = \dfrac{A.100}{6}$ µg C/l/

day, was used where A = the value of the dark uptake of CO_2 by microflora in µg C/l/day.

5.3.2 Estimation technique. The technique of estimation of dark CO_2 uptake (A) is in principle similar to the technique by Steemann Nielsen for the estimation of phytoplankton photosynthesis with the use of $^{14}CO_2$. The only difference is that the experiments have to be made in complete darkness. The radioactivity of the ^{14}C-carbonate used for this estimation also has to be in concentration 5—10 times greater than that used for the estimation of photosynthesis; about 4—$7 \cdot 10^6$ cpm/l in the eutrophic waters, about 8—$10 \cdot 10^6$ cpm/l in the mesotrophic waters and about $20 \cdot 10^6$ cpm/l in the oligotrophic waters. These counts correspond to the bicarbonate-C content of the waters (approximately 12—15 mg C/l). In the waters with less concentration of the salt the amount of ^{14}C has to be proportionally decreased.

Practically, the procedure for the estimation is as follows. The water samples taken with the water bottle are poured into a series of experimental bottles. The bottles have to be previously carefully cleaned, sterilized by acid KI-iodine solution and repeatedly washed with water from the water bottle. Sometimes zooplanktons and the major part of phytoplanktons are removed by previous filtration of the water during pouring it into the experimental bottles (Sorokin, 1965). In this case a special funnel with a membrane filter (pore size, 4µm), or a tube, the end of which is covered with double layers of fine plankton net is attached to the outlet tube of the water bottle.

In case of subsampling of the oxygen deficient waters it is of special importance to avoid the contact of water with air to prevent disturbance of the *in situ* redox conditions.

Bottles with the samples are placed in the dark for 2 hours. Then a portion of ^{14}C-carbonate solution is added under the dark condition. The bottles are exposed at the *in situ* temperature for 24 hours (or longer when working with the cold waters). Then the micro-organisms in the bottles are killed with formaldehyde. The samples are then filtered through membrane filters. After the end of filtration the filters are rinsed with 1—2% HCl directly in the funnel to remove the remaining ^{14}C-bicarbonate. The filters are dried and the radioactivity is counted.

A control estimation should be made in the presence of formaldehyde. The radioactivity of the control filters should be subtracted from the radioactivity of the experimental ones. The value of the assimilation is calculated using the equation by Steemann Nielsen: $A = \dfrac{k \cdot r \cdot 24}{R \cdot T}$ µg C/l, where $K =$ the content of CO_2+bicarbonate-carbon in water in µg C/l; $r =$ the radioactivity of micro-organisms in the sample, cpm; $R =$ total radioactivity of ^{14}C-carbonate added into the sample, cpm, and $T =$ time of incubation in hours.

Methods of estimation of K and R values are described in IBP Handbook No. 12 (1969).

The approximate value of microbial production in sediment (P) can be calculated from data on the dark CO_2 assimilation by benthic microflora (A) µg C/l/day. In completely oxidized sediments (Eh + 400 mv) $P = 10A$ µg C/l; in sediments in which the organic content is high (Eh − 100 mv) $P = 5A$ µg C/l; and in sediments with high content of sulphides (Eh − 200 mv) $P = 2·5A$ µg C/l.

Measurement of dark assimilation of CO_2 in bottom sediments can be made as follows (Sorokin, 1958; Romanenko, 1964b). To 1—2 g of the sediment sample placed in the test tube, 10 ml of water and a portion of ^{14}C-carbonate solution (1—5·10⁵ cpm) are added. The test tube is stoppered with a rubber stopper and the contents mixed. After being held in complete darkness for 1 day the contents of the test tube are fixed with the formaldehyde and transferred into the flask. After a small amount of water is added again, the flask is shaken for 10—15 min. Then a 0·5—2 ml portion of the suspension, after a short period of sedimentation of larger particles, is filtered through a millipore filter. Radioactivity of the filter (r) is counted. In the case of lime sediment samples should be treated preliminarily with 2% HCl. Calculation for the CO_2 assimilation is made by use of the same formula as used for the calculation with water samples.

5.4 Calculation of rate of microbial production from rate of oxygen consumption

Comparison of the rate of oxygen consumption in natural water as measured *in situ* by the oxygen bottles method (cf. 3.1) with the rate of the production of the microbial biomass by the direct microscopical method (cf. 5.2) and by the ^{14}C method (cf. 5.3) indicates that there is a certain relationship between these values. The ratio between the production of biomass of micro-organisms (P mg C/l) and the amount of oxygen consumed (D, mg O_2/l) measured by the direct method is approximately 0·08 (Sorokin, cf. 5.7), and that by the

^{14}C method is 0·086 (Romanenko, cf. 3.5). Therefore from the data on the rate of the oxygen consumption (D, mg O_2/l/day) it is possible to calculate an approximate value of the rate of microbial production by use of the following equation:

$$P = 0{\cdot}08D \ mg \ C/l/day$$

5.5 Production of autotrophic micro-organisms

5.5.1 Estimation of bacterial photosynthesis. The production of bacteria by photosynthesis is measured also by ^{14}C method. The technique used is similar to that used for estimation of dark assimilation (cf. 5.3), except that exposure to light is necessary and that the previous filtration of the sample is not necessary. Since the bacterial photosynthesis takes place only under the strongly anaerobic conditions, the water also must be treated carefully to avoid contact with air during sampling and filling of the subsample bottles.

Bacterial photosynthesis is usually found in a water body where the anaerobic zone exists above the lower boundary of the euphotic zone.

The samples for the measurements of *in situ* photosynthesis by bacteria should be taken between this depth and the upper boundary of the anaerobic zone, at intervals of 0·5—1m. The bottles after the addition of ^{14}C-carbonate are exposed *in situ* at the depths of the sampling. To avoid injury to bacteria by strong light, the experiment should not be done in strong sunshine. A parallel sub-sample bottle is exposed in darkness as control for determining the dark CO_2 assimilation. The latter value has to be subtracted from the value of the total CO_2 assimilation in the light bottles to obtain the real value of the bacterial photosynthesis (Sorokin, 1970).

5.5.2 Estimation of chemosynthesis. Microbial production by chemosynthesis takes place in the layers having contact with the anaerobic zones of the water body, especially in the boundary layers between the anaerobic and aerobic zones. The anaerobic processes of decay of organic matter provide reduced inorganic substances which serve as energy substrates for the chemoautotrophic bacteria. In such circumstances the production of chemoautotrophic bacteria can be regarded as a kind of secondary production. The process of chemosynthesis is of special importance only in the gradient layers of the redox potential (Sorokin, 1964). Outside these layers the significance

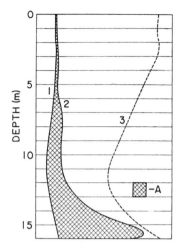

Figure 29. Scheme for calculation of chemosynthetic production in a lake: 1—assimilation of CO_2 by heterotrophic bacteria, calculated from its value measured at the surface and accounting the vertical distribution of heterotrophic bacteria measured by plate count; 2—total dark CO_2 assimilation; 3—plate count of heterotrophs; A—chemosynthesis.

of chemosynthesis in the total bacterial production is very low (Romanenko, 1964).

The chemosynthetic production of bacteria is estimated from the value of the total dark assimilation of CO_2 measured by the ^{14}C method (A) and the value of the CO_2 assimilated by the heterotrophic micro-organisms in the same sample (H). Then the production by chemosynthesis (P_c) is calculated by the equation: $P_c = A - H$. An approximate value of H can be calculated from the equation: $H = 0.06\ P$, where P is the total microbial production estimated by the direct microscopic method. In this case the production by chemosynthesis (P_c) is calculated by the equation: $P_c = A - 0.06\ P$ μg C/l/day where A is total dark assimilation (μg C/l/day), and P is the total production of bacteria (μg C/l/day). The value H is also directly measured at the surface layer of a water body, where the activity of chemosynthetic bacteria is negligible. In this case a vertical profile of the H value in the water body can be obtained, based on the data on the vertical distribution of the heterotrophic bacteria by the plate count, on the assumption that the relative intensity of CO_2 uptake by the heterotrophic bacteria (H) is roughly proportional to their number. Having the distribution of the values H and A it is possible to calculate the CO_2 uptake during chemosynthesis (fig. 29).

5.6 Estimation of biochemical activity of the natural microflora

The localization in water bodies of the specific groups of micro-organisms and the relative intensity of the biochemical processes catalyzed by them are often expressed more directly by their metabolic activities rather than by their cell numbers. This is mainly due to the fact that the proportion of inactive or dormant cells of micro-organisms in natural water is not always constant with the water bodies. The micro-organisms which are actually in the inactive state *in situ* are also counted by the usual microbiological methods. Estimation by the bottle method of the activity of micro-organisms of different physiological types is based upon the measurement of the rate of metabolic reaction after the addition of a specific substrate to water samples in the bottles. The short term incubation of the samples reflects the relative activity of specific group of micro-organisms in the water. As the criterion of the physiological and biochemical activities the BOD, the rate of assimilation of $^{14}CO_2$ or labelled organic compounds are used.

Measurement of the activity of methane- and hydrogen-oxidizing bacteria is carried out as follows: the water sample dispensed into four bottles is aerated by shaking. Using one of these bottles the initial content of oxygen is measured. Into the second and the third bottles small amounts of H_2 and CH_4 are inserted respectively with syringes. The fourth bottle serves as a control one. After exposure for 1—3 days, oxygen content of water in these bottles is measured. As the criterion of activities of methane- and hydrogen-oxidizing bacteria the differences in BOD value between the bottles with CH_4 or H_2 and the control bottle are employed (Kuznetsov 1957).

The metabolic activity of *Thiobacilli* can be determined by measuring the rate of oxidation of thiosulphate added in amount of 10 mg/100 ml to the water samples. The same water sterilized by boiling serves as control. After the incubation, the amount of thiosulphate remained in the water samples is determined by iodometric titration. This value is compared with that in the control bottle. The activity of *Thiobacilli* in the sample is proportional to the rate of oxidation of thiosulphate in it.

The activity of heterotrophic and autotrophic micro-organisms can usually be estimated by determining the $^{14}CO_2$ uptake in water samples, in the presence or absence of the specific substrate. The water sample is poured into the bottle after removal of plankton by filtration. The specific substrate (e.g., H_2, CH_4, S_2O_3, H_2S, NH_4) and the solution of ^{14}C carbonate (0·5—1 10^6 cpm) are added into the control flask.

After the incubation for 0·5—1 day the water is filtered through a millipore filter and the radioactivity of the filter is counted. The activity in this case is expressed as the increase in radioactivity in the bottles in the presence of the specific substrate (Sorokin, 1964, 1970a).

Parsons and Strickland (1962) determined the activity of heterotrophic microflora by adding small amounts of labelled organic substances such as glucose or acetate to the water samples. After incubation the samples are filtered through millipore filters and their radioactivity is counted. The rate of incorporation of ^{14}C into the cells is used in this case as an indicator of the biochemical activity of natural microflora. The measurement of the heterotrophic potential is also used for the marine microflora (Vaccaro and Jannasch, 1966; Sorokin, 1970, 1971).

To measure the heterotrophic potential in water the water sample is poured into 100—300 ml sterilized bottles and a portion of sterilized solution of ^{14}C organic compound is added. As the labelled organic substrate, glucose or protein hydrolysate (cf. 5.7) can be used. Specific activity of the organic substrate should be higher than 1—4 10^6 cpm per 1 mg. The amount of carbon in the solution added into the bottles has to be about 15—20 µg C. The bottles are then incubated at the desired temperature in the darkness. After filtration the filters are rinsed with water and radioactivities of the filters are then counted. The incubation should be stopped when radioactivity of the filters attains 5—10% of the initial activity of substrate added to the bottle. All the samples are fixed with formaldehyde. The samples are filtered through millipore filters (pore size 0·4µm), and washed out as described above.

To measure the heterotrophic potential in bottom sediments (Sorokin, 1969), a 2 ml portion of the sediment sample is placed in a test tube and then 8 ml of sterilized water and a portion of the labelled organic substrate are added. The test tube is stoppered, shaken and incubated in the darkness. After incubation 0·5 ml of 10% HCl is added to the tube. The contents of the test tube are then transferred into a flask and shaken for several minutes. Then 0·5—1 ml of the supernatant is filtered through a millipore filter to estimate radioactivity of the microbial cells. Further procedure is the same as described above.

5.7 The use of continuous culture

The physiological and biochemical activities of micro-organisms in the natural waters can be analysed using continuous culture techniques.

The quantitative determination of the production or disappearance of a chemical substrate over a period of time requires a closed system, *i.e.* sample bottles or other types of culture vessels. As soon, however, as an active microbial population is enclosed in a sample of limited volume, the environmental conditions will change as a corollary of growth. The concentration of the substrates serving as energy source, etc., will decrease and the concentrations of metabolic products will increase. The quantitative composition of species in the mixed population might change quickly, leading to an enrichment culture, especially when a tracer substrate has to be added. These prospective changes of population and activities might not be critical if the time of incubation is short. There is no general rule, however, for a safe margin since microbial activity and population composition will differ from water to water, and so will the response of growth stimulation by an added substrate and of inhibition by an accumulated metabolic product.

Closed culture systems can be avoided by using continuous flow cultures (open systems). In principle, the only technical difference of the latter consists of a constant medium inflow and outflow, and complete mixing in the culture vessel. The theoretical difference of growth in the two systems has been treated and described by Herbert *et al.* (1956). In measuring microbial activities, continuous culture devices offer two major advantages over closed (batch) culture devices: (1) the cultural or environmental conditions can be kept constant for a prolonged period of time, and (2) growth can be measured in the presence of extremely low concentration of the limiting substrate. However, enrichments take place in continuous culture as well as in batch culture, although the selective forces differ (Jannasch, 1965).

In continuous culture enrichments, those organisms can be isolated and investigated that attain the fastest growth under natural conditions, that is, under the extremely low substrate concentrations of the natural environment. Such studies have shown that the common bacteriological media of high nutrient concentrations do not provide optimal growth conditions for those aquatic bacteria that are active under natural conditions (Jannasch, 1967). In other words, the usual plating technique select for micro-organisms that are probably inactive in the natural environment.

In a recent application of continuous culture, growth of a single species is measured in its natural water either in the presence or in the absence of the natural microbial population. In this approach, the rate of growth is calculated from the difference between the experimental dilution rate of the system and the washout rate of the specific organisms tested (Jannasch, 1969). The results show that average generation time of aquatic bacteria range from about 20 hours upwards in oligotrophic waters.

Most applications of continuous culture procedures take advantage of a state at which microbial growth is exactly matched by the washout of cells from the culture vessel. At this 'steady state', at which a continuous culture becomes a 'chemostat', the important parameters of growth (population density, growth rate, concentration of limiting substrate, etc.) may be kept constant for theoretically unlimited time. Steady state cultures make it possible to use simple mathematical treatment of the growth parameters. Such studies require pure cultures and have led to essential information on growth of aquatic bacteria at low substrate concentration (Jannasch, 1967).

As it might be too early in this edition of the Manual to recommend the application of continuous culture for general routine technique, no technical details are given here but may be found in the literature.

5.8 Estimation of efficiency of biosynthesis of microbial cells

The efficiency of the microbial biosynthesis (K_2) is expressed as the percent ratio of the amount of organic matter assimilated by micro-organisms to the total amount of the organic matter involved in the microbial metabolism. This value is estimated usually for microbial cultures using biochemical methods. In most cases, however, the data obtained in cultures grown in rich organic media have been used for the calculations of the efficiency of the processes of microbial productivity in natural waters where the amount and the quality of the organic matter is quite different. For the direct estimation of the efficiency by the natural microflora (K_2) the following methods can be employed:

a. Measurement of K_2 by use of labelled organic matters
As the labelled organic substrate an acid hydrolysate of labelled algae, such as *Chlorella*, can be used. The active algae are resuspended in the medium with very low content of HCO_3^- ion. Then the solution of ^{14}C-carbonate ($10-20.10^6$

cpm/200 ml) is added to it. During the exposure to strong light for several hours the ^{14}C-carbonate added will be almost completely consumed by algae. The algae is harvested by centrifugation or filtration with the membrane filters and then is subjected to acid hydrolysis by 10% HCl at 100° C for 5 hours. Then the hydrolysate is neutralized and filtered through a membrane filter with pore size of 0·1—0·3μm. The hydrolysate is then sterilized in ampoules. The specific radioactivity should be about 3—5·10^{-3} μg C/cpm.

To measure the efficiency a portion of the hydrolysate is added to the water sample. The amount of the hydrolysate to be added is 30—100 μg organic carbon per litre. In this concentration the substrate does not seriously change the initial state of organic matter dissolved in the natural water. The sample after addition of the hydrolysate is incubated in darkness. After each 3—5 hours exposure portions of the water sample are filtered through a millipore filter to measure the radioactivity of the microflora in the water sample. The filters are washed to remove the rest of the labelled dissolved organic matter by subsequent filtration with 2—3 portions of water. The filters are dried and counted. The ratio of radioactivity of micro-organisms in the series (r) to the total radioactivity added to the sample (R) is regarded as the maximum efficiency of utilization of dissolved organic matter by the natural microflora (Sorokin, 1971).

b. Estimation of K_2 by measuring BOD and production rate of micro-organisms
Having known the BOD value of the water sample (D) and the rate of production of the microflora (P) measured by the direct method (cf. 5.2) or the ^{14}C method (cf. 5.3) it is possible to calculate the K_2 value by use of formula
$K_2 = \dfrac{P \cdot 100}{D + P}\%$ when D and P values are expressed as the amount of carbon. An example of the data obtained by this method is given in Table 9.

5.9 Remark on microbial production in freshwater communities

The microbial population plays an important part in the energy flux in the whole community of an aquatic ecosystem. Microbial production in an aquatic ecosystem can be divided into two parts, viz., 'primary' and 'secondary' production. The microbial production which occurs at the expense of allochthonous organic matter (Pa) can be regarded as 'primary' production,

Table 9. Estimation of K_2 value in natural aquatic microflora* (After Y. Sorokin)

No. of experiments	Production of micro-organisms (P)		O_2-uptake (D), µg O_2/l	$\dfrac{D}{P}$	K_2, %**
	CO_2 up-take (A) µg C/l	P, µg biomass /l			
1	1·22	244	210	0·86	24
2	1·30	260	190	0·73	27
3	1·40	280	210	0·75	26
4	3·90	740	590	0·80	25

* Respiratory coefficient is regarded as 1.
** Carbon content in micro-organisms is regarded as 10%.

because it increases the total amount of energy used by the aquatic community.

Microbial production at the expense of the autochthonous sources of energy is regarded as a secondary microbial production (P_s) and is accordingly recognized as a part of the production of the secondary trophic level.

The total annual production of micro-organisms (Pt) can be obtained graphically using the curve of the seasonal changes in the daily production of micro-organisms in the water column and the surface layer of bottom sediments. The secondary microbial production (P_s) can be expressed as that produced at the expense of the organic matter which originates from aquatic plants (and) subjected to microbial decomposition. The latter value (Pd) can be obtained by subtracting from the total production of the aquatic plants (P_p) the part of it which could be consumed by the animal population (P_r): $P_d = P_p - P_r$. The value P_r can be calculated on the basis of feeding ratios of herbivorous zooplankton. The secondary production of micro-organisms is calculated by the equation: $P_s = (P_p - P_r) \cdot 0.25$, provided that the efficiency of bacterial biosynthesis (K_2) is 0·25. Then by subtracting the secondary microbial production from the total microbial production, the 'primary' production of micro-organisms at the expense of the allochthonous organic matter can be calculated: $P_a = P_t - P_s$. The total production of the primary food resources (PP) is calculated as in the sums $P_a + P_p$.

6
Evaluation of the Trophic Role of Micro-organisms

The trophic value of micro-organisms can be examined by (1) growing aquatic animals with micro-organisms as food and (2) measuring the decrease of the concentration of micro-organisms due to the grazing, or measuring the rate of incorporation of carbon or phosphorus of microbial cells into the consumers, using tracer techniques.

6.1 Nonisotopic methods

Determination of the rate of growth of the animals fed with bacterial suspension as a single source of food was reported by Rodina (1948). The consumption of micro-organisms by animals is also estimated by observing the decrease in microbial concentration in bottles (cf. 5.2). This method can be successfully used for the study of grazing of a natural population of micro-organisms by filter feeding animals. The rate of grazing of planktonic bacteria by the natural zooplankton population can be estimated as follows. The water sample is dispensed into two parallel 300—500 ml flasks, with and without filtration through coarse membrane filters (pore size, 5μm). The filtered sample does not contain zooplankton and serves as a control. For the filtration of zooplankton a plankton net of fine mesh can also be used. With both the flasks the initial concentration of micro-organisms is counted by the microscopic method. The flasks are exposed for 6—12 hours *in situ* and after exposure the measurement of microbial concentration is repeated. The rate of grazing can be calculated according the equation described previously (cf. 5.2).

For the study of grazing by specially selected species of animals both flasks are filled with water free of zooplanktons. Then into one of them are put the animals to be tested. The animals used for this experiment should be washed previously with bacteria-free natural water or with 0·04% solution of NaCl, in order to remove the micro-organisms attached to them. For this purpose a simple device can be used (Fig. 30). Another method for washing the animals

Evaluation of the Trophic Role of Micro-organisms

is the consecutive transfer to a series of bottles with bacteria-free water. The estimation of the rate of grazing is made as described before (cf. 5.2). If the multiplication of bacteria is inhibited by the addition of antibiotics, the grazing rate can simply be calculated from the decrease in the bacterial number.

The same experiments can be made on the grazing of micro-organisms grown in cultures. The micro-organisms are washed by centrifugation and put, in a definite concentration, into the bacteria-free water into which then the animals to be tested are placed.

6.2 Isotopic methods

For quantitative study micro-organisms previously labelled with ^{14}C (Sorokin, 1968) or ^{32}P (Rodina and Troshin, 1954) are used as the foods.

^{14}C is more suitable than ^{32}P for obtaining quantitative data. The specific radioactivity of micro-organisms labelled with ^{14}C is more stable as compared with that labelled with ^{32}P. The phosphate binds to the living cells in a very labile state and its uptake and excretion is strongly dependent on the metabolic conditions of the cells.

The principle of the ^{14}C method is as follows. With the suspension of labelled micro-organisms the total number of micro-organisms (N, 10 g/l) and their biomass (B, μg/l) are measured by using direct microscopy (cf. 4.2), and the determination of organic carbon (μg/l), respectively. With the same sample, the radioactivity of micro-organisms is also measured under the standard condition (r, cpm/l). The values $\frac{N}{r}$, $\frac{B}{r}$ or $\frac{C}{r}$ correspond to the reverse specific radioactivity expressed as the number of microbial cells per 1 cpm (N_r), or as the amount of wet biomass (B_r) or carbon of micro-organisms (C_r) per 1 cpm are measured. Thus the number of micro-organisms or their carbon content can be calibrated against the radioactivity. Using the suspension of labelled micro-organisms it is possible to measure the amount of micro-organisms in very low concentration. By estimation of radioactivity (r) in the bodies of animals, which have been fed with the labelled micro-organisms, it is possible to obtain quantitative data on the rate of consumption or assimilation (R) of the food micro-organisms, since the radioactivity can be recalculated to the number of micro-organisms or the biomass of micro-organisms or the amount of carbon of micro-organisms: $R = r \cdot N_r$, $R_2 = r \cdot B_r$, and $R_3 = r \cdot C_r$.

The detailed procedure is as follows:

Heterotrophic bacteria are labelled by growing them in media containing ^{14}C-glucose or ^{14}C-*chlorella* hydrolysate (cf. 5.7). The medium having the following composition per 1 is suitable for this purpose: K_2HPO_4, 0·2 g; NaH_2PO_4, 1 g; $MgSO_4 \cdot 7H_2O$, 0·1 g; $NaNO_3$, 0·2 g; glucose, 0·2 g; Yeast extract, 20 mg. The 200 ml portions of the medium are dispensed into 1 l flasks, and sterilized by autoclaving. The labelled organic substances are sterilized separately by heating at 100°C in the sealed ampoules (cf. 5.7), and then added to the medium. The specific radioactivity of the organic substances to be added is dependent on the purpose of the experiment. Usually the specific radioactivity has to be not less than 10^5 cpm per mg C. Cells of the micro-organisms grown in the culture are harvested and washed by centrifugation and then resuspended in a small portion of water to prepare the cell suspension containing 0·5—1 mg C/ml.

The microflora in the natural water sample, the detritus, or the bottom sediment can also be labelled with ^{14}C by use of the labelled organic substances having high specific radioactivity 5—10 . 10^6 cpm per mg C (Sorokin, 1970). In this case the labelled organic substance is added to the water or to the sediment, in the concentration of 0·1—0·2 mg C/per one litre or 1 g. After holding at 20—25°C for 24 hrs. micro-organisms in the sample incorporate about 40% of the labelled carbon into the cells. The $^{14}CO_2$ formed during the incubation in the suspension has to be removed by acidifying and bubbling of air. After this treatment the suspension is neutralized. In most cases, however, this treatment can be omitted, because the concentration of CO_2 in natural waters is generally very high and the specific activity of $^{14}CO_2$ in the medium is correspondingly too low to have any effect upon the results of the experiments.

For the estimation of the reverse specific activity expressed as N_r (number of cells per 1 cpm) it is necessary to count the number of micro-organisms (N) in the suspension or in the sample by the direct method (cf. 4.2), and also to measure the radioactivity of the cells (r) after being filtered through a membrane filter. The reverse specific radioactivity (N_r) is expressed as $N_r = \dfrac{N}{r}$ cells per 1 cpm. The value expressed as C_r (amount of the organic carbon per 1 cpm) is calculated as the ratio: $C_r = \dfrac{C}{r}$. The organic carbon content in the suspensions is determined by ordinary chemical methods based on wet or dry combustion. The measurement of this value in the labelled microflora of a natural substrate can be done by estimating the wet biomass of bacteria (B) by direct microscopy (cf. 4.2). The specific activity in this case will be: $C_r = \dfrac{B}{10r}$ µg C/cpm.

The radioactivity of carbon in the bodies of small invertebrates such as *Bosmina* and *Rotifers*, can be directly counted on the planchets after washing and drying. For the removal of the labelled micro-organisms from the animal body by washing the device shown in Fig. 30 can be used. The animals

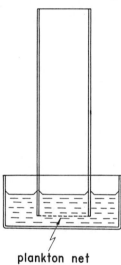

plankton net

Figure 30. A device for washing animals in the feeding experiments.

are then picked up with a pipette, placed onto a drop of 0·02% agar solution in the planchet, and spread inside the standard circle. The preparation is dried and its radioactivity is measured under the standard conditions which must be exactly the same as those during the measurement of N_r or C_r. The average radioactivity per 1 animal is calculated after the measurement. For small animals such as *Bosmina* a correction for self-absorption by the animal bodies is not necessary.

In case of the larger animals the radioactivity can be measured with homogenates of their bodies. A small portion of the homogenate (0·1—0·2 ml) taken with the aid of a calibrated capillary is spread over the surface of the planchet. The preparations are dried and counted by use of the appropriate counter. The correction for selfabsorption is usually also not necessary in this case. The radioactivity found is recalculated per 1 animal body.

More precise value of the radioactivity would be obtained, when ^{14}C in the animal bodies is converted into $^{14}CO_2$ by the combustion and the radioactivity of $^{14}CO_2$ is measured in the state of $Ba^{14}CO_3$ or as $^{14}CO_2$. The latter way also needs a preliminary wet combustion. The resulting $^{14}CO_2$ is collected by vacuum into ion chambers (DC-1000, DC-500, DC-250, DC-100, Nuclear Chicago) and measured with electrometric systems (6010 Dynacon, Nuclear Chicago).

The equipment must be standardized to 100% efficiency by using standard ^{14}C samples (U.S. Bureau of Standards, 1250 cps/g). The use of a liquid scintillation counter is convenient to eliminate selfabsorption problems (Ward et. al. 1970).

The study of the nutritional importance of micro-organisms for aquatic animals is generally made by the following procedures: (a) Evaluation of the significance of micro-organisms, either aggregates or single cells, as source of food, in comparison with the other kind of food, such as algae; (b) Measurement of the rate of filtration; (c) Estimation of the dependence of the rate of feeding upon the concentration of micro-organisms; (d) Estimation of the assimilability of micro-organism; (e) Calculation of the coefficients K_1 and K_2 (the degree of utilization of the consumed or assimilated microbial cells for the growth (Sorokin, 1968; Sorokin, Petipa and Pavlova, 1970; Sorokin, 1971).

The detailed procedure is described in the IBP Manual on the secondary production.

References

ADDANKI S., SOTOS J.F. & REARICK P.D. (1966). Rapid determination of picomole quantities of ATP with a liquid scintillation counter. *Anal. Biochem.* **14**, 261–264.
ALLEN H.L. (1967). Acetate utilization by heterotrophic bacteria in a pond. *Hidrologiai Közlöny.* 7. SZ. **47**, 295–297.
BADGER E.H.M. & PANKHURST E.S. (1960). Experiments on the accuracy of surface drop bacterial counts. *J. Appl. Bacteriol.* **23**, 28–36.
BAIG I.A. and HOPTON J.W. (1969). Psychrophilic properties and the temperature characteristic of growth of bacteria. *J. Bacteriol.* **100**, 845–846.
BAUMANN P. (1968). Isolation of *Acinetobacter* from soil and water. *J. Bacteriol.* **96**, 39–42.
BERGMEYER H.V. (ed.) (1962). Methoden der enzymatischen Analyse. *Verlag Chemie Weinheim, Bergstr.* 1065.
BODE F., HÜBENER H.J., BRÜCKNER & HOERES K. (1952). Eine einfache quantitative Bestimmung von Aminosäuren im Papierchromatogramm. *Naturwiss.* **39**, 524–525.
BROCK T.D. (1967). Mode of filamentous growth of *Leucothrix mucor* in pure culture and in nature, as studied by tritiated thymidine autoradiography. *J. Bacteriol.* **93**, 985–990.
BUCHANAN K.E. & FULMER E.I. (1930). *Physiology and Biochemistry of Bacteria*. Williams & Wilkins Co., Baltimore.
BURRIS R.H. & WILSON P.W. (1957). *Methods in Enzymology*. S.P. Colowick & N.O. Kaplan, (Eds), Academic Press, New York, Vol. 4, 355–366.
CASSELL E.A. (1965). Rapid graphical method for estimating the precision of direct microscopic counting data. *Appl. Microbiol.* **13**, 293–296.
CAVALLI-SFORZA L.L. (1961). Analisi statistica per medici e biologi e analisi del dosaggio biologico. Ed. Boringhieri, Torino.
COCHRAN W.G. (1950). Estimation of bacterial densities by means of the 'Most Probable Number'. *Biometrics*, **6**, 105–116.
COLE, H.A., WIMPENNY J.W.T. & HUGHES D.E. (1967). The ATP pool in *Escherichia coli*. I. Measurement of the pool using a modified luciferase assay. *Biochim. Biophys. Acta*, **143**, 445–453.
COLLINS V.G. (1963). The distribution and ecology of bacteria in freshwater. *Proc. Soc. Water Treatment and Examination*, **12**, 40–72.
COOMBS J., HALICKI P.J., HOLM-HANSEN O. & VOLCANI B.E. (1967). Changes in concentration of nucleoside triphosphates in silicon-starvation synchrony of *Navicula pelliculosa*, *Exptl. Cell Res.* **47**, 315–328.
D'EUSTACHIO A.J. & LEVIN G.V. (1967). Levels of adenosine triphosphate during bacterial growth. *Bacteriol. Proc.* 121.

References

DILWORTH M. (1966). Acetylene reduction by nitrogen-fixing preparations from *Clostridium pasteurianum. Biochim. Biophys. Acta*, **127**, 285–294.

EDMONDSON W.T. & WINBERG G.G. (1971). *A Manual on Methods for the Assessment of Secondary Productivity in Fresh Waters.* IBP Handbook 17. Blackwell Scientific Publications, Oxford.

ENDRES G. & KAUFMAN L. (1937). Die Bestimmung kleinster Mengen von Hydroxylamin, Nitrit und Nitrat. *Ann. Chem.* **530**, 184–194.

FARREL J. & ROSE A.H. (1968). Cold shock in a mesophilic and psychrophilic pseudomonad. *J. gen. Microbiol.* **50**, 429–439.

FROBISHER M. (1962). *Microbiology.* 7th edition. Saunders, Philadelphia & London.

FRUTON J.S. & SIMMONDS S. (1958). *General Biochemistry.* 2nd edition, John Wiley & Sons, Inc., New York. 1977.

GABE D.R., TROSHANOV E.P. & SHERMAN E.E. (1964). Role of micro-organisms in the formation of iron-mangane ores in lakes. *Ac. Sc. U.S.S.R. Leningrad.*

GAK D.S. (1967). The calculation of the bacterial production. *Hydrobiol. J. (Kiev)*, 5 (5).

GAK D.S. (1971). The simplified method of the estimation of the bacterial production. *Hydrobiol. J. (Kiev)*, in press.

GAMBARYAN (1962). On the method of estimation of intensity of the destruction of organic matter in the bottom sediments of the deep water bodies. *Microbiologia*, **31** (5), 895–898.

GOLTERMAN H.L. (1960). Studies on the cycle of elements in fresh water. *Acta Bot. Neerl* 1–58.

GOLTERMAN H.L. (1964). Mineralization of algae under sterile condition or by bacterial breakdown. *Verh. Internat. Verein. Limnol.* **15**, 544–548.

GOLTERMAN H.L. (1969). *Methods for Chemical Analysis of Fresh Waters.* IBP Handbook 8. Blackwell Scientific Publications, Oxford, 172.

GOSSLING B.S. (1958). The loss of viability of bacteria in suspension due to changing the ionic environment. *J. appl. Bacteriol.* **21**, 220–243.

HAMILTON R.D. & HOLM-HANSEN O. (1967). Adenosine triphosphate content of marine bacteria. *Limnol. Oceanogr.* **12**, 319–324.

HANDA N. (1966). Examination on the applicability of the phenol sulfuric acid method on the determination of dissolved carbohydrate in sea water. *J. Oceanogr. Soc. Japan*, **22**, 79–86.

HANDA N. (1967a). The distribution of the dissolved and the particulate carbohydrates in the Kuroshio and its adjacent areas. *J. Oceanogr. Soc. Japan*, **23**, 115–123.

HANDA N. (1967b). Identification of carbohydrates in marine particulate matters and their vertical distribution. *Record Oceanogr. Works Japan*, **9**, 65–73.

HANDA N. (1969). A detailed analysis of carbohydrates in marine particulate matter. *Marine Biology*, **2**, 228–235.

HARDY R.W.F., HOLSTEN R.D., JACKSON E.K. and BURNS R.C. (1968). The acetylene-ethylene assay for N_2 fixation; laboratory and field evaluation. *Plant Physiol.* **43**, 1185–1207.

HAYES F.R. & ANTHONY E.H. (1959). Lake water and sediment: VI. The standing crop of bacteria in lake sediments and its place in the classification of lakes. *Limnol. Oceanogr.* **4**, 299–315.

HAYES F.R. & MCAULAY (1959). Oxygen consumed in water over sediment cores. *Limnol. Oceanogr.* **4**, (3), 291–298.

HENRICI A. & McCOY E. (1938). *Trans. Wis. Acad. Sci.* **31**, 323–361.
HENRICI A.T. (1939). The distribution of bacteria in lakes. *Problems of Lake Biology, Amer. Assoc. Adv. Sci., Washington D.C.*, Publ. No. 10, 39–64.
HERBERT *et al.* (1956). The continuous culture of bacteria: a theoretical and experimental study. *J. Gen. Microbiol.* **14**, 601–622.
HOBBIE J.E. & WRIGHT R.T. (1965a). Bioassay with bacterial uptake kinetics: glucose in fresh water. *Limnol. Oceanogr.* **10**, 471–474.
HOBBIE J.E. & WRIGHT R.T. (1965b). Competition between planktonic bacteria and algae for organic solutes. *Mem. Ist. Ital. Idrobiol.* Suppl., **18**, 175–187.
HOLM-HANSEN O. & BOOTH C.R. (1966). The measurement of adenosine triphosphate in the ocean and its ecological significance. *Limnol. Oceanogr.* **11**, 510–519.
HOLM-HANSEN O (1969). Adenosine triphosphate content of unicellular algae (in preparation).
HWANG J.C. & BURRIS R.H. (1968). Binding sites of nitrogenase. *Federation Proc.* **27**, 639.
IVANOFF M.V. (1955). The method for estimation of bacteria biomass in the water body. *Microbiologia*, **24**, (1), 79–89.
JANNASCH H.W. & JONES G.E. (1959). Bacterial populations in sea water as determined by different methods of enumeration. *Limnol. Oceanogr.* **4**, 128–139.
JANNASCH H.W. (1965). Continuous culture in microbial ecology. *Labor. Practice*, **14**, 1162.
JANNASCH H.W. (1967). Enrichments of aquatic bacteria in continuous culture. *Arch. Microbiol.* **59**, 165.
JANNASCH H.W. (1967). Growth of marine bacteria at limiting concentrations of organic carbon in seawater. *Limnol. Oceanogr.* **12**, 264–271.
JANNASCH H.W. & MADDUX (1967). A note on bacteriological sampling in seawater. *J. Mar. Res.* **25**, 185–189.
JANNASCH H.W. (1968). Competitive elimination of *Enterobacteriaceae* from seawater. *Appl. Microbiol.* **16**, 1616–1618.
JANNASCH H.W. (1969). Estimations of bacterial growth rates in natural waters. *J Bacteriol.* **99**, 156.
KADOTA H., HATA Y. & MIYOSHI H. (1966). A new method for estimating the mineralization activity of lake water and sediment. *Mem. Res. Inst. Food Sci. Kyoto Univ.* **27**, 28–30.
KEPES A. (1963). Permeases: identification and mechanisms. Pages 38–48 *in* Gibbons N.F. (ed), *Recent progress in microbiology*. 8th International Congress for Microbiology, Montreal, 1962. Univ. Toronto Press, Toronto.
KIMBALL J.F. & WOOD E.J.F. (1954). A simple centrifuge for phytoplankton studies. *Bull. Mar. Sci.* **14**, 539–544.
KING W.L. & HURST A. (1963). A note on the survival of some bacteria in different diluents. *J. appl. Bacteriol.* **26**, 504–506.
KOCH B. & EVANS H.J. (1966). Reduction of acetylene to ethylene by soybean root nodules. *Plant Physiol.* **41**, 1748–1750.
KOYAMA T. (1953). Measurement and analysis of gases in sediments. *J. Earth Sci., Nagoya Univ.* **1**, 107–118.
KOYAMA T. (1954). Distribution of carbon and nitrogen in lake muds, *J. Earth Sci., Nagoya Univ.* **2**, 5–14.

KOYAMA T. & TOMINO T. (1967). Decomposition process of organic carbon and nitrogen in lake water. *Geochemical J.* **1**, 109–124.
KOYAMA T. & TOMINO T. (1968). Mineralization process of organic matter in lake water. *Bull. Misaki Mar. Biol. Inst. Kyoto Univ.* **12**, 111–124.
KRISS A.E. (1963). *Marine Microbiology.* (Ed.). Oliver & Boyd, London, 1–536.
KRUSE J.M. & MELLON M.G. (1953). Colorimetric determination of ammonia and cyanate. *Anal. Chem.* **25**, 1188–1192.
KUZNETSOV S.I. (1959). Die Rolle der Microorganismen im Stoffkreislauf der Seen. *VEB. Deutscher Verlag der Wissenschaften Berlin.*
KUZNETSOV S.I. & ROMANENKO W.I. (1963). Microbiological study of the water bodies (Laboratory manual). *Ac. Sci. USSR*, M. L. 127.
KUZNETSOV S.I. & ROMANENKO W.I. (1966). Produktion der Biomass heterotropher Bacterien und die Geschwindigkeit ihrer Vermehrung im Rybinsk-Stausee. *Verh. int. ver. Limnol.* 1603–, 1495–1500. XI. Ecology of Freshwater Organisms. 1. Aquatic Bacteria.
KUZNETSOV S.I. (1969). Microflora of the lakes and its geochemical activity. *Ac. Sci. USSR, Leningrad*, 439.
LYMAN G.E. & DE VICENZO J.P. (1967), Determination of picogram amounts of ATP using the luciferin-luciferase enzyme system. *Anal. Biochem.* **21**, 435–443.
MARGALEF R. (1957). Information theory in ecology. *Gen. System.* **3**, 36–71.
MELCHIORRI-SANTOLINI U. (1966). Pelagic heterotrophic bacteria in the Ligurian Sea and Lago Maggiore. *Mem. Ist. Ital. Idrobiol.* **20**, 261–287.
MELCHIORRI-SANTOLINI U. (1969). I conteggi colturali della microflora batterica degli ambienti limnici. *Atti XV Congr. Soc. Ital. Microbiol.,* Torino-Saint Vincent, Ottobre 1969.
MEYNELL G.G. & MEYNELL E. (1965). *Theory and Practice in Experimental Bacteriology.* Cambridge University Press, Cambridge.
MILES, A.A. & MISRA S.S. (1938). The estimation of bactericidal power of the blood. *J. Hyg.* **38**, 732–749.
MOORE S. & STEIN W.H. (1951). Chromatography of amino acids on sulfonated polystyrene resins. *J. Biol. Chem.* **192,** 663–681.
MOSS M.L. (1942). Colorimetric determination of iron with 2,2'-bi-pyridyl and with 2,2',2"-tripyridyl. *Ind. Eng. Chem., Anal. Chem. Ed.* **14**, 862.
MOZEN M.M. & BURRIS R.H. (1954). The incorporation of ^{15}N-labelled nitrous oxide by nitrogen fixing agents. *Biochim. Biophys. Acta,* **14**, 577–578.
MUNK W.H. & RILEY G.A. (1952). Absorption of nutrients by aquatic plants. *J. Mar. Res.* **11**, 215–240.
NIKITIN D.I. (1964). The use of the electron microscopy for the study of the soil suspension and the cultures of micro-organisms. *Pochvovedenie,* N6, 86–91.
NIKITIN D.I. et al. (1966). *New and Rare Forms of Soil Bacteria.* Nauka, Moscow.
NISKIN (1962). *Deep-Sea Res.* **9**, 501–503.
OANA S. (1957). Bestimmung des Argon im besonderen Hinblick auf gelöste Gase in natürlichen Wässern. *J. Earth Sci., Nagoya Univ.* **5**, 103–124.
PARSONS T.R. & STRICKLAND J.D.H. (1962). On the production of particulate organic carbon by heterotrophic processes in sea water. *Deep-Sea Res.* **8**, 211–222.
PERFILIEV B.V. & GABE D.R. (1961). *The Capillary Method for Micro-organisms Investigations.* Nauka, Moscow.

PERFILIEV B.V. & GABE D.R. (1964). The role of micro-organisms in the formation of iron-manganese deposits.
PETO (1953). A dose-response equation for the invasion of micro-organisms. *Biometrics*, **9**, 320–335.
POMEROY L.R. & JOHANNES R.E. (1966). Total plankton respiration. *Deep-Sea Res.*, **8**, 211–222.
POSTGATE J.R. & HUNTER J.R. (1962). The survival of starved bacteria. *J. gen. Microbiol.* **29**, 233.
POSTGATE J.R. & HUNTER J.R. (1964). Accelerated death of *Aerobacter aerogenes* starved in the presence of growth-limiting substrates. *J. gen. Microbiol.* **34**, 459–473.
RASUMOV A.S. (1932). *Mikrobiologiya*, **1**, 131–146.
RASUMOV A S. (1947). Methods of microbiological studies of water. Moscow. VODGEO.
RIPPEL-BLADES A. (1947). *Grundriss der Mikrobiologie*. Springer Verlag, Berlin.
RODINA A.G. (1948). The role of bacteria and yeasts in the nutrition of *Daphnia magna*. *Trans. Zool. Inst. Ac. Sci. USSR*, 8(3).
RODINA A.G. & TROSHIN A.S. (1954). The use of radioactive indicators in the study of the nutrition of aquatic animals. *Doklady Acad. Sci. USSR*, **98**, 297–299.
RODINA A.G. (1963). Microbiology of detritus of lakes. *Limnol. Oceanogr.* **8**, 388–393.
RODINA A.G. (1965). Methods of the water microbiology (Manual). *Science, M., L.* 1–362.
ROMANENKO W.I. (1963). The potential ability of the microflora in water to the heterotrophic CO_2 assimilation and to the chemosynthesis. *Microbiologia*, **32** (3), 668–674.
ROMANENKO W.I. (1964). Heterotrophic assimilation of CO_2 by the aquatic microflora. *Microbiologia*, 33 (4), 679–683.
ROMANENKO W.I. (1964a). The potential ability of the bottom microflora to the heterotrophic assimilation of CO_2 and to the chemosynthesis. *Microbiologia*, **33** (1), 134–139.
ROMANENKO W.I. (1965). The relation between the consumption of oxygen and CO_2 by heterotrophic bacteria during the growth in the presence of peptone. *Microbiologia*, **34** (3), 391–396.
ROMANENKO W.I. (1965). The ratio between the oxygen consumption and CO_2 uptake by the heterotrophic bacteria during their growth in the peptone media. *Microbiologia*, **34** (3), 397–402.
ROMANENKO W.I. & ROMANENKO V.A. (1969). The destruction of the organic matter in the bottom sediments of the Rybinsk reservoir. *Trans. Inst. Inland Waters, Acad. Sci. USSR*, **19** (22), 24–31.
ROMANENKO W.I. (1969). The time of generation and the time of the doubling of the CO^2 assimilation by heterotrophic bacteria. *Information Bulletin of the Inst. of Inland Waters, Acad. Sci. USSR*, No. 4, 8–11.
ROMANOVA A.P. (1958). Intensity of the development of bacterial flora in the shallow water of Baikal Lake (by the slide method). *Microbiologia*, **27** (5), 634–639.
ROMANOVA A.P. & ZONOV A.I. (1964). On the estimation of production of bacterial biomass in the water body. *Doklady Acad. Sci. USSR*, **155** (1), 194–197.
RUCHTI J. & KUNKLER D. (1960). Enzymatische Bestimmung von Glucose, Fructose und Saccharose in Gewässern. *Schweiz. Ztschr. f. Hydrol.* **28**, 62–68.

SCHÖLLHORN R. & BURRIS R.H. (1966). Study of intermediates in nitrogen fixation. *Fed. Proc.* **25**, 710.
SCHÖLLHORN R. & BURRIS R.H. (1967). Acetylene as a competitive inhibitor of N_2 fixation. *Proc. Natl. Acad. Sci. U.S.* **58**, 213–216.
SIEBURTH Y.M. (1963). A simple form of ZoBell bacteriological sampler for shallow water. *Limnol. Oceanogr.* **8**, (1), 489–493.
SLOGER C. & SILVER W.S. (1967). Biological reductions catalyzed by symbiotic nitrogen-fixing tissues. *Bact. Proc.* 112.
SOROKIN Y.I. (1958). Studies of chemosynthesis in bottom sediments with the use of ^{14}C. *Microbiologia*, **27**, 206–213.
SOROKIN Y.I. (1960). Bacteriological water bottle. *Bull. Inst. Biol. Reservoirs. Acad. Sci. USSR*, **6**, 53–54.
SOROKIN Y.I. (1964). A quantitative study of the microflora in the central Pacific Ocean. *J. du Conseil*, **29**, 25–40.
SOROKIN Y.I. (1964). On the trophic role of chemosynthesis in water bodies. *Int. Rev. Ges. Hydrobiol.* **49** (2), 307–324.
SOROKIN Y.I. (1965). On the trophic role of chemosynthesis and bacterial biosynthesis in water bodies. *Mem. Ist. Ital. Idrobiol.* **18** (Suppl.), 187–205.
SOROKIN Y.I. (1968). The use of ^{14}C in the study of nutrition of aquatic animals. *Mitt. Intern. Ass. Theoretical and Appl. Limnol.* **16**, 1–41.
SOROKIN Y I. (1969). On the estimation of activity of heterotrophic microflora in the ocean with the use of labelled organic matter. *Microbiologia*, **38** (5), 868–872.
SOROKIN Y.I. (1970). On the estimation of activity of heterotrophic microflora in the ocean with the use of labelled organic matter. *Microbiologia*, **39** (1), 149–155.
SOROKIN Y.I. (1970). Interrelations between sulphur and carbon turn-over in meromictic lakes. *Arch. f. Hydrobiol.* **66**, 391–446.
SOROKIN Y.I., PETIPA T.S. & PAVLOVA E. (1970). Quantitative study of the nutritional importance of marine bacterial plankton. *Oceanologyia*, **10**, 332–340.
SOROKIN Y.I. (1971). On the role of bacteria in the productivity of tropical oceanic waters. *Int. Rev. Ges. Hydrol.* **56**, 1–48.
STEFACNOV. S.B. (1962). The simple method of the preparing of the preparation of the crude virus suspension for the electron microscopy. *Biophysica*, **76**, (6), 725–726.
STEWART W.D.P., FITZGERALD G.P. & BURRIS R.H. (1967). In situ studies on N_2 fixation using the acetylene reduction technique. *Proc. Natl. Acad. Sci. U.S.* **58**, 2071–2078.
STEWART W.D.P., FITZGERALD G.P. & BURRIS R.H. (1968). Acetylene reduction by nitrogen-fixing blue-green algae. *Arch. Mikrobiol.* **62**, 336–348.
STROKES J.L. & REDMOND M.L. (1966). Quantitative ecology of psychrophilic microorganisms. *Appl. Microbiol.* **14**, 74–78.
STRANGE R.E. & DARK F.A. (1962). Effect of chilling on *Aerobacter aerogenes* in aqueous suspension. *J. gen. Microbiol.* **29**, 719–730.
STRASKRABOVA V. (1968). *Limnologica*, **6**, 29–36.
STREHLER B.L. (1968). Bioluminescence essay: principles and practice. In: *Methods of Biochemical Analysis* (D. Glick, ed., Interscience Publishers, N.Y.), **16**, 99–181.
STRICKLAND J.D.H. & PARSONS T.R. (1968). A practical Handbook of Seawater Analysis. *Bull. Fisheries Res. Board Can.* No. 167, 245–249.

SUGAWARA K. (1939). Chemical studies in lake metabolism (1). *J. Chem. Soc., Japan*, **14**, 375–451.
TANAKA M. (1951). Colorimetric microdetermination of manganese in the field. *J. Chem. Soc. Japan, Pure Chem. Section*, **22**, 29–34.
TEZUKA Y. (1968). A method for estimation of bacterial respiration in natural water. *Japan J. Ecol.* **18**, 60–65.
TROITSKY A.S. & SOROKIN Y.I. (1957). On the methods of the calculation of the bacterial biomass in water bodies. *Trans. Inst. Biol. Inland Waters, Acad. Sci. USSR*, **19**, 85–90.
VACCARO R.F. & JANNASCH H.W. (1966). Studies on heterotrophic activity in sea water based on glucose assimilation. *Limnol. Oceanogr.* **11**, 596–607.
VOLLENWEIDER R. (1969). *A Manual on Methods for Measuring Primary Production in Aquatic Environments*. IBP Handbook 12. Blackwell Scientific Publications, Oxford, 213.
WARD E.Y., WONG B. & ROBINSON I. (1970). A liquid scintillation procedure for determining the effect of size on selfabsorption of ^{14}C in *Daphnia pulex*. *Limnol. Oceanogr.* **15** (11), 645–652.
WILLINGHAM & BUCK (1966). *Deep-Sea Res.* **12**, 693–695.
WILSON P.W. & UMBREIT W.W. (1937). Mechanism of symbiotic nitrogen fixation: III. Hydrogen as a specific inhibitor. *Arch. Mikrobiol.* **8**, 440–457.
WOOD E.J.F. (1962). A method for phytoplankton study. *Limnol. Oceanogr.* **7**, 32.

Index

Acetate 22
  uranyl 52
Acetic
  Lugol's acid solution 22
Acetylene
  reduction method 3
Achromobacter 63
Acid
  amino 37
  fatty 24
  Lugol's acetic, solution 22
  osmic 52
  succinic 27
Acinetobacter 63
Activated sludge
  in sewage treatment 73
Activity
  biochemical, of the natural microflora 88
  gas-producing 29
Adenosine triphosphate 3, 71
  concentration of 73
  determination of 74
  molecule of 74
Aerobic
  decomposition, estimation of 25
  mineralization of organic substances 25
Affinity
  substrate 22
Agar Medium
  nutrient 60
Aggregates
  bacterial 69
Aggregation
  level of planktonic bacteria 76
Algae
  blue-green nitrogen-fixing 5
Allochthonous
  organic matter 1, 92

Amino Acids
  proteinous 37
Amylacetate 51
Anaerobic
  decomposition of organic matter 24, 83
  zones 83
Analysis
  chromatographic 30
  sanitary, of water and sewage 63
Animals
  filter-feeding 94
Antibiotics 95
Argon 35
Assimilability
  of micro-organisms 98
Assimilation of
  dark, of CO_2 83
  heterotrophic CO_2 23
ATP 3, 71, 73
  determination of 74
  molecule of 74
Autochthonous
  sources of energy 93
Autolysis 24
Autotrophic
  CO_2 77
  micro-organisms 86

Bacteria
  chemoautotrophic 23
  dead 50
  heterotrophic, production of 82, 83
  hydrogen oxidizing 88
  living 50
  methane oxidizing 88
  planktonic, aggregation level of 76
  psychrophilic 64
  respiration rate of 17

Index

Bacterial
  aggregates 69
  photosynthesis 86–87
Batch culture 90
Bioassay 23, 73
Biochemical
  activity of natural microflora 88
Bioluminescent reaction 73
Biomass
  calculation of 48
  determination of 49
  estimation 73
  of micro-organisms 71
  indicator 73
Biometer
  of the DuPont Co. 76
Biosynthesis
  of microbial cells 91
Bipyridyl
  2-2¹, method 36
BOD 92
Bosmina 97
Bottles
  evacuated 40
  sampler 40
  serum 4
  vaccine 4
  Van Dorn 73
Buffer
  Tris 27, 74
Bulbs
  rubber 40
Burette
  gas 30

Cadmium
  copper reduction method 36
  nitrate 27
Calanoida 76
Calculation
  of biomass 48
Capillary
  glass cells 57
  method 53
  peloscope 53, 57
  periphytonometer 53
  sealed glass 40
Carbohydrate
  dissolved 36

particulate 36
Carbon
  mineralized 36
Carbonate ^{14}C 84
Carbon dioxide $^{14}CO_2$
  dark assimilation of 83
  dark uptake of 83
  heterotrophic assimilation of 23
  heterotrophic uptake of 23
Catabolic powers 63
Cells
  dormant 88
  size 48
  viable, of micro-organisms 59
Chemoautotrophic
  bacteria 23
Chemoautotrophs 83
'Chemostat' 91
Chemosynthesis
  estimation of 87
  production by 87
Coefficient
  respiratory 26
Chlorella 91
  hydrolysate ^{14}C 96
Chloroplasts 50
Chromatographic analysis 30
Chromatography of paper 37
Cold shock 61
Collodium solution 51
Concentration
  natural substrate 22
Condenser
  quartz 50
Constant
  Michaelis-Menten 21
  transport 21
Consumption
  oxygen 86

Contaminated waters 45
Copper net planchets 51
Corer tube 27
Counters
  scintillation 75
  scintillation (liquid) 19
Counting
  direct microscopic, of micro-organisms 44

Counts
 microscopic 70
 plate 59, 63, 71
Crowding 61
Culture
 batch 90
 continuous 90
 continuous, devices 90
 elective 63
 enrichment 90
 selective 63
 steady state 91

Dark assimilation of CO_2 83
 uptake of CO_2 83
Dead bacteria 50
Death
 'substrate-accelerated' 61
Decomposition 15
 aerobic, estimation of 25
 anaerobic, of organic matter 24, 83
Dehydrogenase activity in bottom sediments 27
Denitrification
 microbial processes of 2
Denitrified nitrogen 34
Detectors
 light 75
Detrital material 71
Detritus 73
 organic 50
DHA 27
Differential staining 44
Diffusion
 simple 21
Diluent fluids 61
Dilution
 ratio 65
 series (MPN) 64
Diphenyloxazole 2·5 (PPO) 19
Direct microscopic counting of micro-organisms 44
Dissolved carbohydrate
 determination of 36
Distribution-size
 of living cells and detrital particles 71
Dithionite 3
Dormant cells
 of micro-organisms in natural water 88

Eadie enzyme Kinetics 22
Elective culture 63
Electrodes
 oxygen 17
Electron microscope, use of 51
Energy
 autochthonous sources of 93
Enrichment culture 90
Enzyme system, N_2 fixing 7
Enzyme-substrate-complex 20
Error
 standard 69
 statistical 62
Erythrosin 45, 60
Ethanolamine 19
Ethylcellosolve 19
Euphotic zone 87
Evacuated bottles 40
Extracts, Lyophilized water 74

Fatty acids 24
Ferredoxin 3
Filter-feeding animals 94
Filtration, rate of 98
Firefly
 lanterns (*Photinus pyralis*) 74
 tails 73
Fixation
 biological N_2 3
 measurements of N_2 3
 nitrogen 7
Fixed nitrogen 34
Fluorescence antibody method 51
Fluorescent microscopy 50
Flushing 6
Forces, selective 90
Formaldehyde 51
Formazan 27
Fractionation, solvent 37
Funnel, glass 44

Gas
 -producing activity 29
 burette 30
 microbial evolution of 33
Generation
 of C_2H_2 9
Geochemical methods 33

Glass
  capillary cells 57
  funnel 44
Glucose 22
  oxidizing activities 19
Glutaraldehyde 50
Grazing 94
Growth
  inhibition by metabolic products 62
  optimal conditions for aquatic bacteria 90

Halophilic micro-organisms 53
'Heat attachment'
  of microbial cells 53
Herbivorous zooplankton
  feeding ratios of 93
Heterotrophic
  bacteria, production of 82, 83
  CO_2 77
  CO_2 assimilation 23
  micro-organisms 63
  potential, measurement of 89
  uptake of carbon dioxide $^{14}CO_2$ 23
Heterotrophs 83
Hirsch funnel 6
Homogenity, statistical 46
Homogenized mud samples 30
Hydrogen
  molecular 24
  oxidizing bacteria 88

Incubation
  condition of 63
  temperature 64
Indicator, biomass 73
Inhibition, growth 62
Inorganic sulphur compounds
  oxidation of 2
Invertebrates, small 97
Iodometric titration 88
Ionization detector 11
Iris-Mud-Sampler 29
Isotopic methods 95

Kinetics
  enzyme, Eadie 22
  Michaelis-Menten 20, 22

Labelled
  organic matter 91
  substrates 22
Lanterns
  firefly (*Photinus pyralis*) 74
Level, aggregation
  of planktonic bacteria 76
Light
  emission 73
  detectors 75
Liquid scintillation counter 19
Lineweaver-Burk enzyme Kinetics 22
Living bacteria 50
Luciferase 74
Luciferin, reduced 74
Lugol's acetic acid solution 22
Lyophilized water extracts 74

Mass spectrometer 12
Maximum uptake velocity 22
Mesotrophic waters 45
Metabolic products 62, 90
Methane CH_4 11
  evolution of 27
  oxidizing bacteria 88
Methylcellosolve 38
Michaelis-Menten
  enzyme kinetics 20
  constant 21
Microbial
  cells (biosynthesis of) 91
  gas evolution 33
  numbers 49
  production 1, 86
Microbiological samplers 40
Micro-organisms
  assimilability of 98
  autotrophic 86
  benthic 71
  halophylic 53
  heterotrophic 1, 63
  periphytonic 71
  planktonic 71
  trophic value of 94
Microscope, electron 51
Microscopic counting, direct 44, 70
Microscopy, fluorescent 50
Microzones 57
Microzonal distribution 57

Mineralization
  activity 19
  aerobic, of organic substances 25
  carbon 36
  nitrogen 36
  of nitrogenous organic matter 23
  of organic matter 15, 18
  processes of 2
Molecular hydrogen 24
Molecule of ATP 74
Monosaccharides 36
  composition of 37
Moraxella 63
Most Probable Number 69
Moulds 2
Mud
  homogenized samples 30
  Iris, Sampler 29
  syringe samples 27

Natural substrate concentration 22
Net, copper planchets 51
Ninhydrin 38
Niskin sampler 42
Nitrate, Cadmium
Nitrification 2
Nitrogen
  denitrified (or fixed) 34
  fixing enzyme system 7
  mineralized 36
  reservoir for 10
Nitrogenous organic matter
  mineralization of 23
Nonisotopic methods 94
Number—Most Probable 69
Nutrient
  agar medium 60
  regeneration 2

Oligosaccharides 37
Oligotrophic waters 45
Optimal growth conditions
  for aquatic bacteria 90
Organic matter
  allochthonous 92
  anaerobic decomposition of 24, 83
  detritus 50
  labelled 91

  mineralization of nitrogenous 23
  mineralization rate of 18
Organisms
  periphytonic 53
  photosynthetic 50
  psychrophilic 61
Osmic acid 52
Oxidation of inorganic sulphur compounds
  2
Oxide, silver 36
Oxidized sediments 85
Oxygen 74
  consumption 16, 86
  electrodes 17
Oxidizing bacteria
  hydrogen 88
  methane 88

Paper chromatography 37
Particulate carbohydrate 36
  matter 61, 73
Peloscope, capillary 53, 57
Periphytonic
  micro-organisms 71
  organisms 53
Periphytonometer, capillary 53
Phenolphthalein 26
Phenol sulphuric acid method 36
Phenyloxazolyl, benzene (POPOP) 19
Phosphorus wolframate 52
Photinus pyralis (firefly lantern) 74
Photosynthesis
  bacterial 86–87
  phytoplankton 84
Photosynthetic organisms 50
Phytoplankton photosynthesis 84
Planchets, copper net 51
Planktonic
  bacteria, aggregation level of 76
  micro-organisms 71
Plastic bags 40
Plate
  count 59, 63, 71
  trapping 55
Plating medium 63
Polysaccharides 36
Potential, heterotrophic 89
'Poured-plate' technique 60
Power, catabolic 63

Production
  by chemosynthesis 87
  microbial 1, 86
  microbial photosynthetic 1
  microbial secondary 1
  of heterotrophic bacteria 82, 83
  'primary' 92
  rate of 92
  'secondary' 92
Producers—secondary 71
Products, metabolic 62, 90
Proteinous amino acids 37
Pseudomonas 63
Psychrophilic
  bacteria 64
  organisms 61
Pyridine-pyrazolone method 36

Quartz condenser 50

Rate
  of production 92
  washout 91
  of filtration 98
Reaction, bioluminescent 73
Redox-potential 27
Reduced luciferin 74
Reduction
  acetylene 3
  Cadmium-copper 36
  C_2H_2 5
  sulphate 2
Reservoir for $^{15}N_2$ 10
Respiration rate of bacteria 17
Respiratory coefficient 26
Rotifers 97
Rubber bulbs 40

Sampler
  bottle 40
  Iris-Mud 29
  microbiological 40
  Niskin 42
  Syringe-type 42
Sampling
  of sediments 43
  techniques 40

Sanitary analysis of water and sewage 63
Scintillation counter 75
  counter (liquid) 19
Sealed glass capillary 40
Secondary
  producers 71
  production 92
Sediment
  oxidized 85
  sampling 43
Selective
  culture 63
  forces 90
Serum bottles 4
Sewage 63
  treatment 73
Shock, cold 61
Silver oxide 36
Simple-diffusion 21
Size distribution of living cells and detrital particles 71
Sludge, activated 73
Solvent fractionation of particulate carbohydrate 37
Spectrometer, mass 12
'Spread-plate' technique 59
Staining, differential 44
Standard error 69
Statistical
  error 62
  homogeneity 46
'Steady state' 91
  cultures 91
Streptomyces 2
Subsamples 43
Substrates
  'accelerated' death 61
  affinity 22
  concentration 22
  labelled 22
  low concentration 91
Succinic acid 27
Sulphate reduction 2
Sulphur compounds
  oxidation of inorganic 2
Sulphuric acid, Phenol method 36
'Surface drop' technique 59
Syringe 40
  mud samples 27
  sampler 42

Temperature, incubation 64
Tetrachlorethylene 27
Thiobacilli 88
Titration, iodometric 88
Tracer techniques 16, 94
Transport constant 21
Trapping plates 55
Treatment of sewage 73
Triphenyltetrazolium chloride 27
Tris-
  buffer 27, 74
  oximethylaminomethane 27
Trophic
  level 45
  value of micro-organisms 94
Turnover time 22

Uptake
  dark, of CO_2 83
  heterotrophic, of $^{14}CO_2$ 23
  maximum velocity 22
  system 21
  velocity of 20
Uranyl acetate 52

Vaccine bottles 4
Van Dorn bottle 73

Velocity
  maximum uptake 22
  of uptake 20
Viable cells of micro-organisms 59
Viruses 2
VUFS 44

Washout rate 91
Water
  column 25
  extracts lyophilized 74
Waters
  contaminated 45
  mesotrophic 45
  oligotrophic 45
Winkler's method 17
Wolframate, phosphorous 52

Yeasts 2

Zones
  anaerobic 83
  euphotic 87
Zooplankton, herbivorous 93